Innovation on Demand

This book describes a revolutionary methodology for enhancing technological innovation called TRIZ. The TRIZ methodology is increasingly being adopted by leading corporations around the world to enhance their competitive position. The authors explain how the TRIZ methodology harnesses creative principles extracted from thousands of successful patented inventions to help you find better, more innovative solutions to your own design problems. You'll learn how to use TRIZ tools for conceptual development of novel technologies, products and manufacturing processes. In particular, you'll find out how to develop breakthrough, compromise-free design solutions and how to reliably identify next generation products and technologies. This explains the book title. Whether you're trying to make a better beer can, find a new way to package microchips or reduce the number of parts in a lawnmower engine, this book can help. Written for practicing engineers, product managers, technology managers and engineering students.

Victor R. Fey is Adjunct Professor of Mechanical Engineering at Wayne State University and a Principal Partner and co-founder of the TRIZ Group. He is one of the few TRIZ Masters personally certified by Genrikh Altshuller, the creator of TRIZ. He has authored or co-authored seven patents, over 30 papers and two books.

Eugene I. Rivin is a professor and director of the Machine Tool Laboratory in the Mechanical Engineering department at Wayne State University and a co-principal and co-founder of the TRIZ Group providing TRIZ training and methodology for problem solving and technological forecasting. He is a Fellow of ASME and SME and an active (full) member of CIRP. Rivin holds over 60 patents and has authored more than 160 papers and 15 books.

Innovation
on Demand

Victor Fey and Eugene Rivin

CAMBRIDGE
UNIVERSITY PRESS

CAMBRIDGE UNIVERSITY PRESS
Cambridge, New York, Melbourne, Madrid, Cape Town, Singapore, São Paulo

Cambridge University Press
The Edinburgh Building, Cambridge CB2 2RU, UK
Published in the United States of America by Cambridge University Press, New York

www.cambridge.org
Information on this title: www.cambridge.org/9780521826204

First published 2005

Printed in the United Kingdom at the University Press, Cambridge

A catalog record for this publication is available from the British Library

Library of Congress Cataloging in Publication data

ISBN-13 978-0-521-82620-4 hardback
ISBN-10 0-521-82620-9 paperback

Contents

Preface

This book describes a powerful methodology for systematic technological innovation called TRIZ (pronounced *treez*). TRIZ is the Russian acronym for the Theory of Inventive Problem Solving (*Teoriya Resheniya Izobretatelskikh Zadach*). The book acquaints the reader with basic tools of TRIZ and their applications to the conceptual development of novel technologies, products, and manufacturing processes.

This book is principally intended for practicing engineers whose responsibilities run the gamut from R&D activities, to design, to shop floor administration. Engineering students will also benefit from its contents. The book describes the vital role of TRIZ in the process of technological innovation. Technology managers who use TRIZ approaches often find strategic opportunities that non-users tend to overlook. They capitalize on these opportunities by developing new products and processes, as well as novel services and organizational structures.

TRIZ originated in the former Soviet Union, where it was developed by Genrikh Altshuller and his school, beginning in 1946. TRIZ was used extensively in the Soviet space and defense industries to enable engineers to overcome difficult technological challenges within an inefficient economic system. It was virtually unknown in the West, however, until a translation of one book by Altshuller was published in 1984 (*Creativity as an Exact Science*, by Gordon and Breach, New York). While the book initiated a few devotees to TRIZ, a poor translation minimized its impact.

In 1991, a TRIZ-based software package, developed by the Invention Machine Corporation, was demonstrated in New York and commercially launched. Although the software attracted significant interest, it was, essentially, only a series of illustrated problem-solving analogies that failed to reveal the thought process required for the effective application of the methodology itself. Since the software's users were by and large unfamiliar with the thought processes behind TRIZ, they were unable to fully utilize the power of this methodology.

Throughout the 1990s, small consulting groups began to appear in the West, usually founded by immigrants from the former Soviet Union. Their principals were experienced TRIZ practitioners. Many of those pioneers were students and collaborators of TRIZ's founder, Genrikh Altshuller. Those groups were solving problems for their client companies and training customers in TRIZ fundamentals. As a result of these efforts, leading corporations in the U.S. and overseas have reported significant benefits from using TRIZ (e.g., see Raskin, 2003).

In 1993, after modifying the approach to TRIZ training adopted in the Soviet Union to better suit American audiences and after offering several successful public seminars, the authors started a four-credit course on TRIZ at Wayne State University in Detroit. Students appreciated the course immediately. The experiences accumulated through the course instruction, as well as knowledge gained from training and consulting projects for industrial clients in the US and overseas (through The TRIZ Group founded by the authors), convinced us that a comprehensive text-book on TRIZ principles was necessary.

This book covers the basic concepts and tools of contemporary TRIZ. The only criterion for including a subject was whether it was essential for the successful application of the methodology. Every notion, method and technique is illustrated by real-life examples gleaned from different areas of technology. Many examples are based on our own inventions, both patented and otherwise, made with the help of TRIZ. Most chapters end with a set of problems and exercises that give the reader an opportunity to sharpen his or her understanding of the earlier described material.

Today's TRIZ contains numerous problem analysis and concept generation tools, not all of them well formalized. In this book, we primarily describe the most formalized tools, such as the ideality tactics, the separation principles, the sufield analysis, the Standards, ARIZ, and some others. Major non-formalized, but still powerful tools of TRIZ, such as the system operator, the size–time–cost method, and the "smart little people" method are described in other books (e.g., Altshuller, 1994, 1999; Mann, 2002).

This book addresses the application of TRIZ to two basic activities which engineers and scientists in a technology-based company may be responsible for: (A) The improvement of existing technologies, products and manufacturing processes ("problem solving"); and (B) The development of next-generation technologies, products and processes ("technological forecasting"). TRIZ has proven to be greatly beneficial for both these activities.

Proliferation of TRIZ in the West started from marketing Invention Machine™ software. Now there are several software products on the market that claim computerization of TRIZ. Usually, these products contain vast knowledge bases of various TRIZ techniques and physical effects, as well as some problem formulation tools. They also contain libraries of good design concepts gleaned from various engineering domains. In our experience, engineers trained in TRIZ to the degree of deep understanding of its principles and thought processes, can be helped by these products. This can be compared with other computer-assisted engineering activities. Extremely powerful software packages for finite element analysis (FEA) can treat a huge variety of stress/strain problems. The results, however, can be right or wrong depending on the model of the analyzed problem that was constructed by FEA software's user. So, one must be good in strength of materials and/or theory of elasticity to generate adequate FEA results. The same can be said about CAD packages. Although these give many prompts to the operator, it is hard to imagine somebody without a mechanical (or electrical) design background and education making a good set of design drawings.

The book is organized in six chapters and five appendixes. Chapter 1 is an introduction that demonstrates the need for systematic innovation. Chapter 2 describes TRIZ tools for resolving conflicts between competing design requirements. Application of these tools results in compromise-free design solutions. Chapter 3 introduces a TRIZ substance–field approach used for modeling physical interactions in technological systems. Chapter 4 describes the Algorithm for Inventive Problem Solving, the most powerful problem analysis and concept development tool of TRIZ. Chapter 5 describes the foundation of TRIZ – the laws of technological system evolution. Chapter 6 outlines a comprehensive approach for guided development of next-generation products and manufacturing processes based on the laws of evolution.

Appendix 1 contains a brief biography of Genrikh Altshuller, the creator of TRIZ. Appendix 2 outlines an alternative TRIZ approach to resolving conflicts. Appendix 3 expands upon the subject of the substance–field modeling approach, introduced in Chapter 3, and describes the 76 Standard Approaches to Solving Problems and the algorithm for using these Standards. Appendix 4 presents an overview of the application of TRIZ to resolving management (i.e., non-engineering) challenges. Appendix 5 contains a glossary of TRIZ terms.

For a comprehensive understanding of contemporary TRIZ, reading of the whole book is desirable. As with many textbooks, however, a selective reading may satisfy some individuals not so much concerned with applications of TRIZ to specific problems but rather interested in assessing its overall potential. Those readers who are mostly interested in the problem-solving tools of TRIZ, should pay more attention to Chapters 2–4. The readers with the primary interest in the management of technology innovation and development of next-generation technologies will mostly benefit from Chapters 5 and 6.

We know that this first TRIZ textbook is not perfect and are grateful in advance for any constructive comments and criticisms.

1   Introduction

Technology is as old as mankind. Myriad small and large innovations have shaped the world, and are molding the future of civilization. The prevailing majority of these innovations have been developed haphazardly; their creators have not used any organized approach to finding new ideas. Despite the great past achievements of a random approach to innovation (the wheel, the automobile, the radio, the airplane, the computer, antibiotics, to name just a few), that approach has become increasingly inefficient in today's fiercely competitive marketplace. This chapter shows the principal shortcomings of *random innovation*, and the need to replace it with a method of *systematic innovation*.

1.1   Product development process

Every new product – whether the "product" is a technology, a device or a production process – originates from a new concept. To become a product, a concept must be generated, then evaluated and, finally, developed. This flow of activities constitutes the *product development process* (*PDP*) (Fig. 1.1).

The process begins with the recognition of a *need*. Then, the designer must transform this need into a *clearly defined problem*, or a set of problems. The output of this stage is a problem(s) statement accompanied by a list of various constraints (e.g., performance specifications, manufacturing limitations, economic conditions, statutory restrictions, etc.).

In the next phase, various *conceptual solutions* to the problem(s) are generated. Here, the most important *decisions* are made which bring together engineering, production, and commercial aspects of the problem.

In the following phase of the process, the generated *concepts* are evaluated against various criteria, and the most promising ones are selected for designing a *prototype*. The prototype is then built and *tested*. During this process, necessary corrections are usually made to the conceptual solution.

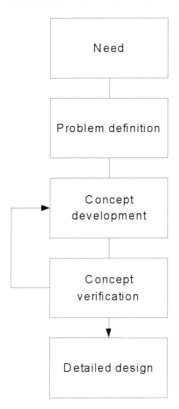

Fig. 1.1. Phases of a typical product development process.

Finally, in the *detail design phase*, the design is fully developed and the dimensions of all the components are determined, the materials specified, etc. Then, detailed drawings are produced.

Modern engineering sciences possess an extensive arsenal of powerful analytical methods and software tools for the efficient evaluation of new concepts, and the development of these concepts into successful processes or product lines. This arsenal has a centuries-long history of gradual evolution.

In the past, there was no other way to test an attractive idea, than to practically try it. Almost all basic chemical compounds have been obtained through a tremendous amount of blind testing by alchemists. Even Thomas Edison, while working on the design for the incandescent bulb, would perform over 6000 experiments with a vast variety of materials before settling on a satisfactory filament.

As science and technology continue to develop, the boundary between what is feasible and what is not becomes better understood. Many products, processes and their environments can, today, be reasonably well simulated. Nowadays, the chemical industry brings numerous new compounds to market every year. This could not be possible without well-developed theoretical methods for rational analysis and synthesis of proposed formulations. While working on a problem, an engineer can filter out weak concepts by using knowledge obtained from basic education, from his or her experiences and those of his or her predecessors, from information in patent and technical literature and, of course, by using computer-based systems such as computer-aided design (CAD), computer-aided manufacturing (CAM), finite element analysis (FEA), computational fluid dynamics (CFD) and others that substantially facilitate the product development process.

1.2   Stumbling blocks in PDP

The least understood and, therefore, often poorly managed first two phases of the product development process are *identification of a need* and *concept development*.

1.2.1   First stumbling block: technology strategy

The first phase – need identification – is, in addition to the market research, also the process of defining a technology strategy. The key question here is (or should be): "*What is the next winning technology to satisfy the potential or perceived market need?*" Answering this question requires a very good understanding of the trends of technology evolution.

In today's fiercely competitive market, manufacturing companies have to gamble their future on the market's next wide acceptance of a product or a technology innovation. They have to figure out what this "winning" innovation will be so as to better allocate adequate financial backing, manpower and other resources. Mistakes in making predictions result in yielding the advantage to competitors.

In 1990, U.S. Steel, then a leader in the American steel market, had a choice of investing in either the conventional hot-rolling technology or in a new compact strip production (CSP) technology. U.S. Steel decided to further improve the well-established, hot-rolling technology. Their competitor, Nucor Steel, funded the development of CSP. Today, Nucor is the leading US steel producer while its formerly formidable competitor is largely marginalized in the marketplace (Cusumano & Markides, 2001).

The two giants of film-based photographic equipment, Kodak and Polaroid, did not recognize the emergence of digital imaging technologies. They each received brutal blows from competitors that pioneered products employing those technologies (Leifer *et al.*, 2000).

Chester Carlson, the creator of xerography, offered his invention to dozens of companies. Each and every one of them turned him down, thus missing out on one of the most successful business opportunities of the twentieth century.

In 1992, start-up Palm, Inc. developed the prototype for its Palm Pilot. The gadget, however, did not excite venture capitalists in Silicon Valley. They saw it as just another personal digital assistant (similar to Hewlett-Packard's Psion, Apple's Newton, and

Windows CE PCs) and refused to fund bringing it to the market (Penzias, 1997). Unable to raise money on their own, Palm became a subsidiary of U.S. Robotics in 1995. Shortly thereafter, the Palm Pilot became hugely successful. Today Palm is again an independent and thriving company offering an impressive array of state-of-the-art handhelds and accessories.

The list of companies that lost their competitive position to more innovative rivals can be easily extended, but so can be the list of successful, "winning" products and technologies that did not have the initial support of management, or of the financial community.

Many business publications and consultants praise market research as the most reliable way to assess the market viability of emerging innovations. While various market research techniques often prove to be very useful for incremental improvements, they often mislead when used to appraise breakthrough innovations.

In the late 1960s, Corning became interested in a promising low-loss optical fiber technology. The company consulted with AT&T – an obvious potential customer – as to the prospects for this technology. AT&T guesstimated that a noticeable need for optical fibers would arise only in the first decade of the twenty-first century. Corning, however, was much more optimistic and set off on the development of this technology. The company had its major commercial success with MCI in the early 1980s (Lynn *et al.*, 1997).

The inability of even the best experts to predict the future of technology is illustrated by the following often-quoted phrases.

- "Heavier than air flying machines are impossible," Lord Kelvin, President of the Royal Society, 1885.
- "Everything that can be invented has been invented," Charles H. Duell, Director of the US Patent Office, 1899.
- "Who the hell wants to hear actors talk?" Harry Warner, Warner Brothers, 1927.
- "There is a world market for maybe five computers," Thomas Watson, Chairman of IBM, 1943.

A more recent example: Bill Gates could not foresee the rise of the Internet.

Hindsight is always 20/20, but how can we make sure that the products or technologies being pursued now are the ones that the market will need? How can we choose the best solutions? What are the criteria that will allow us to select the most promising concepts? Conventional approaches to the identification of next-generation winning products and technologies cannot provide reliable answers to these questions. The TRIZ perspective on these questions is addressed in Chapters 5 and 6.

1.2.2   Second stumbling block: concept development

Today, just as for many centuries, novel ideas in engineering (as well as in other areas of human activity) are mainly produced by the *trial-and-error* method. The essence of this method is a persistent generation of various ideas for solving a problem. There are

Fig. 1.2. Absent glass cannot break.

no rules for idea generation, and the process is often stochastic. If an idea is weak, it is discarded and a new idea is suggested. The flow of ideas is uncontrollable, and attempts (trials) are repeated as many times as is needed to find a solution.

Although seemingly random, most trials have a common attribute: they are more numerous along a so-called vector of *psychological inertia*. This inertia is determined by our cultural and educational backgrounds, previous experiences and "common sense." More often than not, psychological inertia is created by a deceptively innocuous question, *"How?"* (e.g., "how to fix this problem") that nudges the problem-solver toward traditional approaches, dims the imagination and is a key hurdle on the track to the best solution. In fact, the best solution often lies elsewhere, in territories that our common sense deems useless.

Example 1.1

A classic example relates to the psychological inertia of "rocket scientists," supposedly the most educated, sophisticated and innovative of all engineers. In the 1970s, the United States and the Soviet Union fiercely competed in the area of Moon exploration. Unable to afford the costs of an Apollo-like project for landing astronauts on the Moon, the Soviet Union decided to launch an unmanned lunar probe to deliver an autonomous vehicle to the back (dark) side of Moon's surface. The vehicle was to transmit TV pictures of that not-yet seen side to Earth. The incandescent light bulb was found to be the best light source, however existing light bulbs would not survive the impact of landing on the lunar surface. Even the most durable bulbs, those used in military tanks, cracked at the joint between the glass and the socket during tests.

A major effort was started to figure out *how* to strengthen the glass bulb. The situation was reported to the director of the whole Moon-landing project. His first question was: *"What* is the purpose of the bulb?" The answer was obvious – to seal a vacuum around the filament. However, there is an abundance of perfect vacuum on the Moon! This simple question solved the problem – no glass bulb was needed (Fig. 1.2).

Table 1.1. *Muffler alternatives*

Type of muffler	Cost ($)	Noise level (dB)	Back pressure (psi)
A	60	79	7.65
B	90	79	7.40
C	60	82	7.40
Targets	60	79	7.40

Example 1.2

A team of seasoned automobile designers was developing an exhaust system for a new truck. Specifications contained various performance and economic criteria, the most important ones being the noise level, back pressure (affecting the engine's efficiency) and cost. All of these attributes are largely influenced by the specific design of the muffler. There was no commercially available muffler that would meet all the target criteria, so for several months (!), the design team brainstormed the solution space, searching for an adequate muffler design.

Various solutions on *how* to improve the muffler designs were proposed, but none complied with the targets. Three alternative muffler designs A, B, and C that offered the parameters closest to the specifications were identified (Table 1.1). Only the very expensive muffler B had an acceptable technical performance, but its cost was excessive by a large margin. The new truck had to be launched, and budget-breaking muffler B was grudgingly approved.

Relief came unexpectedly from two young co-eds who, at the time, were taking the TRIZ class at Wayne State University. Analysis of the situation by TRIZ methods lead them to question *what* had to be achieved. The existing noise regulations address not the system components (muffler), but the noise exposure of a person on a sidewalk (Fig. 1.3a). Accordingly, the co-eds suggested adopting muffler C, but modifying another part of the exhaust system so as to meet the noise-level specifications. The solution was very simple and elegant.

While the original exhaust system was equipped with a straight tail pipe, essentially aiming towards the microphone, the co-eds bent the end of the pipe downward (Fig. 1.3b); now the sound, produced by the exhaust, does not affect the by-stander (and the measuring microphone), but is deflected to the ground.

Experiments showed that this solution met all the required criteria, and it was immediately implemented.

The fact that such simple solutions in both examples were missed, illustrates the obstructing power of psychological inertia.

The trial-and-error method results in valuable time being wasted when searching for solutions to difficult problems. Damages from the hit-and-miss approach are associated

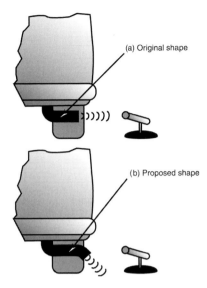

Fig. 1.3. Exhaust system modification.

with lost competitiveness and a waste of manpower and financial resources. Nor does it help that the random generation and selection of concepts fails to provide for the experience gained by solving one problem, and utilizing that experience to solve other problems.

Accordingly, the need for improving the concept generation process has long been necessary (some pre-TRIZ methods for creativity enhancement are described in Appendix 2).

1.3   TRIZ

The *Theory of Inventive Problem Solving* (*TRIZ*), developed by Genrikh Altshuller (Altshuller and Shapiro, 1956; Altshuller, 1994, 1999), states that while the evolution of technology is apparently composed of haphazard steps, in the long run it follows repeatable patterns. These patterns can be applied to the systematic development of technologies – both to solving product design and production problems and to the development of next-generation technologies and products.

TRIZ deals not with real mechanisms, machines and processes, but with their models. Its concepts and tools are not tied to specific objects and, therefore, can be applied to the analysis and synthesis of any technology regardless of its nature. TRIZ also treats all products, manufacturing processes and technologies as *technological systems* (more on that in Chapter 2).

The premise of TRIZ is that *the evolution of technological systems is not random, but is governed by certain laws.*

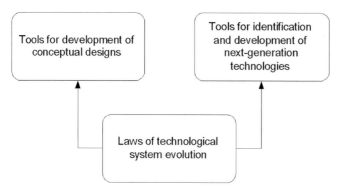

Fig. 1.4. Structure of TRIZ.

Knowledge of these laws allows the anticipation of the most likely next steps that the evolution of a technology will take. It also helps design better systems faster, without wasting time and resources on a random search for solutions.

One can draw an analogy between the use of the laws of technological system evolution (laws of evolution, for short) and the laws of mechanics. If the position of a moving object is known for a certain moment of time, its future position can be found by solving the corresponding equations of motion. The laws of evolution serve as "soft equations" describing the system's "life curve" in the evolution space. If the current system configuration is given, the future configurations can be reliably "calculated" (forecasted).

The name, *Theory of Inventive Problem Solving*, reflects Altshuller's initial intention when developing TRIZ. He wanted to replace the uncertainty of the trial and error process with a structured approach to resolving difficult engineering problems. However, the logic of research led him to develop a methodology that far exceeds the needs of immediate problem solving and allows for prediction of future challenges and, often, for their resolution.

Contemporary TRIZ is both a theory of technological evolution and a methodology for the effective development of new technological systems. It has two major subsystems based on the laws of technological system evolution: a set of methods for developing conceptual system designs, and a set of tools for the identification and development of next-generation technologies (Fig. 1.4).

Logically, the next chapter should describe the laws of evolution, since they constitute the foundation of TRIZ. From a pedagogic perspective, however, it is more beneficial to first introduce the concept development tools of TRIZ. To resolve this dilemma, the presentation of the laws of evolution is split into two parts. In this chapter, the laws are only briefly summarized, so that the reader may refer to them while learning the problem analysis and concept development tools (Chapters 2–4). A comprehensive discussion of the laws and their uses is given in Chapters 5 and 6.

1.3.1   Overview of the laws of evolution

The following laws of evolution have been formulated to date:
- law of increasing degree of ideality
- law of non-uniform evolution of subsystems
- law of transition to a higher-level system
- law of increasing dynamism (flexibility)
- law of transition to micro-level
- law of completeness
- law of shortening of energy flow path
- law of increasing substance–field interactions
- law of harmonization of rhythms.

The *law of increasing degree of ideality* states that *technological systems evolve toward an increasing degree of ideality*. The degree of ideality is essentially the benefit-to-cost ratio. The capabilities of various products (e.g., cell phones, computers, cars) are endlessly increasing, while their prices fall. Systems with a higher degree of ideality have much better chances to survive the long-run market selection process, i.e., to dominate the market.

The *law of non-uniform evolution of subsystems* states that *various parts of a system evolve at non-uniform rates*. This creates *system conflicts*, which offer opportunities for innovation. When improving a system by conventional means, one system's attribute is usually improved at the expense of deteriorating another attribute (e.g., the enhancement of dynamic characteristics of a car by equipping it with a more powerful engine at the expense of increased fuel consumption). Such a situation is called a *system conflict*. While conventional design philosophy urges the designer to seek the least expensive compromise, Altshuller found numerous methods for overcoming system conflicts, i.e., for satisfying both conflicting requirements.

Technological systems evolve in a direction from mono-systems to bi- or poly-systems (the *law of transition to higher-level systems*). Systems usually originate as mono-systems designed to perform a single function, and then gradually acquire capabilities to perform more, or more complex, functions. For example, a pencil (mono-system) → pencil with eraser (bi-system) → set of pencils of various hardness, various colors, etc. (poly-system). Eventually, the bi- and poly-systems may merge into a higher-level mono-system performing more complex functions.

According to the *law of increasing dynamism (flexibility)* rigid structures evolve into flexible and adaptive ones. A new system is usually created to operate in a specific environment. Such a system demonstrates the feasibility of the main design concept, but its applications and performance parameters are limited. In the course of evolution, the system becomes more adaptable to the changing environment.

The *law of transition from macro- to micro-level* states that *technological systems evolve toward an increasing fragmentation of their components*. In other words, shafts, wheels, bearings and other "macro" components are gradually replaced with molecules, ions, electrons, etc. that can be controlled by various energy fields. The evolution of cutting tools is an example of such a transition: from metallic cutters to electrochemical machining to lasers.

The *law of completeness* states that an autonomous technological system has four principal parts: working means, transmission, engine and control means. At early stages of the development of technological systems, people often perform the tasks of some of these principal parts. As the systems evolve, the "human components" are gradually eliminated. In a system "photographic camera", an object to control is the light, a working means is a set of lenses, and a focus-adjusting mechanism is a transmission. In manual cameras, the user's hand serves as both the engine and control means. Automatic cameras use an electric motor and a microprocessor to adjust the focus.

Technological systems evolve toward shortening the distance between the energy sources and working means (the *law of shortening of energy flow path*). This can be a reduction of energy transformation stages between different forms of energy (e.g., thermal to mechanical) or of energy parameters (e.g., mechanical reducers, electrical transformers). Transmission of information also follows this trend.

As its name suggests, the *law of increasing controllability* captures the evolution of controlled interactions among the systems' elements.

The *law of harmonization* states that *a necessary condition for the existence of an effective technological system is coordination of the periodicity of actions (or natural frequencies) of its parts.* The most viable systems are characterized by such a coordination of their principal parts when actions of these parts support each other and are synchronized with each other. This is very similar to production systems, in which "just-in-time" interactions between the principal parts of the manufacturing sequence result in the most effective operation. If the synchronization principle is violated, then the system's components would interfere with each other and its performance would suffer.

1.4   Summary

Today's market demands unique innovations that provide a conspicuous value to the customer. Companies must meet that challenge with fewer resources, at the lowest cost, with higher quality, and with shorter design cycle times. Conventional approaches to such unconventional demands simply will not get the job done. Systematic innovation in products and processes is an imperative for competitive leverage. Such innovation is possible only if the approach to its generation is equally unconventional. TRIZ is such an approach. It does not rely on psychological factors; it is based on the discovery of repeatable, and therefore predictable, patterns in the evolution of all technological systems (laws of technological system evolution). These patterns help engineers solve complex product design and production problems efficiently and economically.

2 Resolving system conflicts

Problem solving is the core of concept development. Solving problems usually starts with attempts to find solutions by using conventional means. Frequently, however, such an approach does not work. Suppose, for example, that we want to enhance the comfort of a car. One of the simplest known ways to achieve this goal is to enlarge the passenger compartment. This will result in increased weight and size (drag resistance), and consequently in increased fuel consumption. Thus, gaining advantage in one part of the car will be negated by disadvantages in other parts.

An engineering design problem becomes difficult when attempts to meet one constraint make another constraint, or constraints, unattainable. Designers are usually taught that it is impossible to have the "best of both worlds," so they take for granted that trade-offs are unavoidable.

There is not a universal approach to distinguish the best trade-off from a variety of potential options. Much effort is spent on trade-off studies, but rarely can the engineer claim that the selected trade-off is the best one possible, and that the entire design space was examined. As a rule, many products and processes are designed with less than optimal performance.

Another, more fundamental, concern is the very nature of trade-offs, which is an incomplete fulfillment of opposing requirements. In other words, trade-offs are inherently suboptimal designs.

This chapter describes concepts and tools of TRIZ that make developing trade-off-free – i.e., breakthrough – designs possible.

2.1 Technological systems

All entities – biological, societal, technological, and others – can be viewed as *systems*. This means that, at any level of decomposition, an entity is made up of interacting parts and each of these parts, in turn, consists of smaller parts, until the level of "elementary particles" is reached (a limit determined by the extent of modern scientific knowledge).

Any technological system – be it a pencil, book, beverage, vehicle, space station, or an assembly line – is designed and built to perform *functions*.

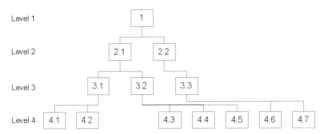

Fig. 2.1. Typical hierarchical (multi-level) system.

Technological systems are organized as *hierarchies*. In a hierarchy, any system contains subordinate subsystems, and itself may serve as a component for a higher-level system (e.g., subsystems 3.1 and 3.2 in system 2.1; Fig. 2.1). For example, a microprocessor is a subsystem of the "mother board," which is a part of a higher-level subsystem of "computer," which in its turn is a part of the system of "computer network." On the other hand, the microprocessor consists of several layers of different materials whose electric and mechanical characteristics are determined by their designs and/or by their molecular structures, which in turn are determined by the properties of the constituent atoms, and so on. Properties of each subsystem are influenced by the properties of the higher- and lower-level systems.

One system can belong to various higher-level systems that may impose conflicting requirements on this system. For example, a watch exists in at least three domains – technical (timekeeping), societal (conveying the owner's status), and in the domain of human physiology. For the last, it should be compatible with human anatomy and bodily processes. Some people are allergic to metals. When gold touches their skin, they experience an allergic reaction, such as eczema. This unpleasant reaction may present a challenge for designers of high-end watches.

Many systems originate as *proto-systems* ("crowds"), with their elements completely independent or loosely connected. The removal of any element from the "crowd" has little or no effect on the other elements, or on the whole structure. A technological system emerges when *functional connections* begin to develop among these elements. Migration from proto-systems to systems is one of the principal vectors of technological evolution (Fig. 2.2).

Example 2.1

In a conventional kitchen, appliances (stove, microwave oven, refrigerator, dishwasher, etc.) are independent. In a "smart kitchen," now being aggressively developed by many appliance and telecom companies, these appliances are networked via a home computer.

The network enables the user to perform a variety of functions: monitoring food inventory, buying groceries online, etc. Additionally, the smart kitchen can be linked to non-kitchen appliances, such as a computer, TV, telephone, air conditioner, water

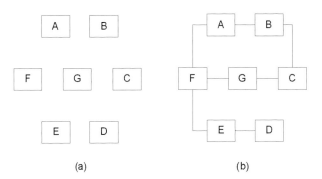

Fig. 2.2. Transition from a "crowd" (a) to a system (b).

heater, etc. and used for obtaining information about weather and road conditions or sending messages via the Internet. Some network-enabled kitchen appliances have already been introduced to the market.

Example 2.2

A similar shift toward increasing interactions between discrete elements is taking place in the transportation area. To make road traffic more safe and efficient, the Japanese Ministry of International Trade and Industry has been developing so-called "cooperative driving" using inter-vehicle communication. The inter-vehicle communication systems use infrared and ultrasound sensors to measure the distance between moving vehicles and their respective speeds. An onboard computer uses this data to tell the driver the best time to merge into another lane. The computer also rapidly halts the vehicle in the case of an abrupt stop by the vehicle in front of it. Preliminary tests of the inter-vehicle communication systems show their high potential utility (Tsugawa, 2000).

Any system has certain characteristics making it not equal to a simple summation of its constitutive elements (otherwise, there would be no purpose for the system). Philosophers have known this for a long time; Aristotle stated that the whole is greater than the sum of its parts. Therefore, the main advantage of a system over separate elements is its *systemic effect*. The emergence of a systemic effect is a reliable criterion that the design is right. Consider, for example, a car with added active suspension. Such a car endows the passengers not only with a more comfortable ride (which is the initial reason for adding the active suspension), but it also has a much higher effective speed since it does not require slowing down at bumps and potholes.

In some cases, a system exhibits properties that are actually opposite to those of its parts. Ingredients of a 'gin-and-tonic' cocktail are gin (usually bitter, with an unpleasant alcoholic taste), tonic water (bitter quinine solution), lime (extremely acerbic taste, much stronger than an acidic lemon), and ice. The combination of these not very appealing elements is a cocktail that is very pleasant and tasty for many people. It is refreshing and, at the same time, a relatively strong alcoholic beverage.

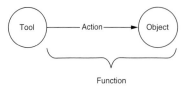

Fig. 2.3. Model of a function.

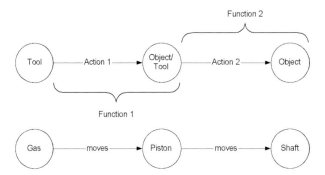

Fig. 2.4. Duality of tools and objects.

2.2   Functions

Technological systems exist to perform functions. When a customer buys a system (i.e., a product or process), he or she essentially pays for the functions performed by this system.

A function involves two components. One of these components, *an object*, is to be controlled, e.g., moved, weighed, magnetized, split, painted, detected, measured, etc. Since an object cannot normally control itself, it needs another component, *a tool*, to execute the required operation. A description of a function includes the *action* produced by both the tool and the object (action receiver) (Fig. 2.3). *Nouns* are used for describing objects and tools, and *verbs* are used to describe actions. For example, "The seat *supports* the body", or "A cup *holds* water," or "The valve *stops* the flow."

The following definitions are used in function descriptions.
- A function is an intended direct physical action of the tool on the object.
- An object is a component to be controlled.
- A tool is a component that directly controls the physical parameters (behavior) of the object.

According to these definitions, in the process of driving nails with a hammer, the hammer head, rather than the whole hammer, is a tool (because it directly affects the object – a nail).

The tool in one "tool–object" pair can be the object in another pair, and vice versa (Fig. 2.4).

Table 2.1. *Examples of correct and incorrect (in the TRIZ context) formulations of functions*

Common formulation	Correct formulation
Hot air *dries hair*	Hot air *evaporates water*
Fan *cools the body*	Fan *moves the air (gas)*
Lens *magnifies* the object	Lens *diffracts light*
Incandescent bulb *illuminates the room*	Incandescent bulb *emits light*
Diode *rectifies an electric current*	Diode *blocks an electric current of a certain polarity*
Lightning rod *attracts lightning*	Lightning rod *conducts electric current*
Windshield *protects the driver*	Windshield *stops the elements*

Function descriptions may sound counter-intuitive at times. For example, one would commonly assume that the function of a ship propeller is to drive the ship. However, this would be an inaccurate formulation, because it violates the definition of a function. There is no direct physical action of the propeller onto the ship. The propeller blades move (push) the water; this makes the ship move. Thus, the function of the propeller is to move the water.

Another common inaccuracy in defining a function is to use *non-physical* terms. These are usually technical parameters (e.g., temperature, pressure, productivity, etc.), consumer-related characteristics (e.g., quality, reliability, accessibility, etc.), and phenomena (e.g., friction, etc.). "The air conditioner improves the tenants' comfort" is an incorrect function formulation when practicing TRIZ. Improved comfort is the result of cooling the air by the air conditioner. Table 2.1 contains other examples of correct and incorrect (in the TRIZ context) function formulations.

Various tools may be used to generate the same action. In order to drive (move) a nail into the wall, for instance, one can use a hammer, pliers, a brick, or a shoe heel.

A technological system usually performs one or more *primary functions* (PF) and one or more *auxiliary functions* (AF). For example, the primary function of the roof of a car is to protect the passengers and the passenger compartment from the elements (rain, snow, dust, etc.). The roof can also support a luggage rack; this is its auxiliary function. A combined printer/fax/copier unit carries out three primary functions.

The PF is usually supported by one or more AFs. Correspondingly, tools can be classified as *main tools* (MT) and *auxiliary tools* (AT). Table 2.2 shows an example of ranking the functions performed by a hammer.

Auxiliary tools can be further classified as *enabling, enhancing, measuring,* and *correcting.*

The enabling auxiliary tools (e.g., a handle of the hammer) support the performance of the primary function. The enhancing auxiliary tools, as the name suggests, boost (or modify) the performance of the main tools (e.g., the arms of a lounger increase its comfort). Similarly, the measuring auxiliary tools gauge or detect the parameters of a

Table 2.2. *Functions and tools associated with driving nails with a hammer*

Object	Nail
Primary function	Driving (moving) the nail
Main tool	Hammer head
Auxiliary function	Moving the hammer head
Auxiliary tool	Handle

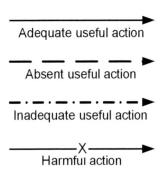

Adequate useful action

Absent useful action

Inadequate useful action

Harmful action

Fig. 2.5. Useful and harmful actions.

system's components (e.g., a control strip on a battery indicates the charge level), while correcting auxiliary tools alleviate or eliminate undesirable effects in the system (e.g., thermal insulation in a thermos).

The practical role this classification plays in TRIZ problem analysis and solution processes is described further in this chapter.

2.2.1   Actions

Actions, performed by tools, can be viewed either as *useful (desirable)* or as *harmful (undesirable)*. This nomenclature is inherently subjective, because the same action can be regarded differently, depending on one's point of view. Piercing the armor with a bullet can please the shooter, but will most likely upset the person protected by the armor.

TRIZ considers four types of useful and harmful actions that are graphically indicated in Fig. 2.5.

(1) Adequate useful action (may benefit from further enhancement).

(2) Inadequate useful action (requires enhancement).

(3) Absent useful function (requires introduction).

(4) Harmful action (requires elimination).

This chapter describes operations with adequate useful and harmful actions. Operations with the other types of actions will be explained in Chapter 3.

2.3   Ideal technological system

Since technological systems are designed and built to perform certain functions, a better system obviously requires less material to build and maintain, and less energy to operate, to perform these functions. The concept of an *ideal technological system* was introduced by Altshuller; it is a system whose mass, dimensions, cost, energy consumption, etc. are approaching zero, but whose capability to perform the specified function is not diminishing.

The ideal system does not exist as a physical entity while its function is fully performed.

The concept of the ideal system allows one to concentrate on the *function to be performed* (i.e., to ask the question, **What** *is needed?*) rather than on improving the existing system, which is performing the required function (asking the question, **How** *can the system be fixed?*). This seemingly minor distinction helps to dramatically reduce the psychological inertia, and assists in developing breakthrough solutions to formidable problems (see Examples 1.1, 1.2, and Problems 2.2–2.5, 2.10).

Functions must be performed by some material entities; it would be wrong, however, to conclude that ideal systems cannot exist in the real world. There are many cases in which the functions of one system are delegated to another system that was initially designed to perform other functions. This means that the function of the ideal (non-existent) system is performed by some other, already present, system.

Example 2.3

Ice cream is often sold in plastic or paper cups. These cups are disposed of after performing their primary function (containing a scoop of ice cream). A more ideal packaging means (system) is a waffle or "sugar cone" cup. Such a cup simultaneously performs two functions: enhancing the taste and holding the ice cream.

It is useful to introduce the concept of *degree of ideality* that represents the *benefit-to-cost ratio* of the system, or the ratio of its functionality, to the sum of various costs associated with the building and functioning of the system (Fig. 2.6).

In this *qualitative* formula, the functionality is assessed by the number of functions performed by the system, and by the level of performance of these functions. The costs can be expressed in various quantifiable units, such as dollars, or measures of size, weight, or number of parts, etc. The problems can relate to measurable attributes

$$\text{Degree of Ideality} = \frac{\text{Functionality}}{\text{Costs} + \text{Problems}}$$

Fig. 2.6. Qualitative expression of degree of ideality.

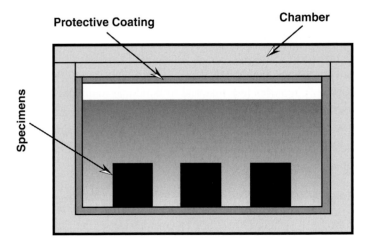

Fig. 2.7. Testing alloys.

(e.g., level of noise, humidity, etc.) as well as to unquantifiable, even intangible ones (e.g., difficulty of handling, discomfort, lack of sense of privacy, etc.).

2.4   Mini- and maxi-problems

At the beginning of the problem-solving process, one usually faces an *initial situation* associated with certain disadvantages that should be eliminated. These disadvantages may be eliminated either by changing a given system by altering one of its subsystems, or by modifying some higher-level system. Thus, various problems associated with the initial situation can be formulated.

Problem 2.1
The corrosive effects of acids on metal alloys are studied in a chamber (Fig. 2.7). The chamber is filled with an acid, closed and various combinations of temperature and pressure are created inside. To protect the chamber walls from destruction, they are lined with a corrosion-resistant glass. When an intense vibration was applied to the chamber, the glass cracked, thus exposing the chamber walls.

This situation can be resolved by addressing the following problems.
- How to improve the protective coating?
- How to make the chamber walls corrosion-resistant?
- How to apply vibrations to the specimens without affecting the coating?
- How to simulate the interaction between the acid and the specimens, without conducting actual tests?

TRIZ divides all problems into *mini-problems* and *maxi-problems.* A mini-problem is when *the system remains unchanged or is simplified, but its shortcomings disappear or a desirable improvement is achieved.* A maxi-problem does not impose any constraints on the possible changes; in this case, a new system, based on a novel principle of functioning, can be developed (often at a much greater cost).

The problem-solving process should be started by addressing a mini-problem, since its solution will require minimal resources and thus might be more economical and easier to implement.

2.5   System conflicts

Improving a system's attributes may lead to the deterioration of its other attributes. Collisions such as weight *vs.* strength, power *vs.* fuel consumption or productivity *vs.* space, etc. are called *system conflicts.*

A system conflict is present when
- a useful action simultaneously causes a harmful effect, or
- the introduction (intensification) of the useful action or the elimination (alleviation) of the harmful action causes an inadequacy, or an unacceptable complication of either one part or of the whole system.

The solution of a problem associated with a system conflict can be attempted either by finding a trade-off between the contradictory demands, or by satisfying all of them. Trade-offs (compromises) do not eliminate system conflicts, but rather soften them, thus retaining harmful (undesirable) actions or shortcomings in the system. Conflicting requirements keep "stretching" the system and, over time, grow increasingly incompatible. When the time comes to further advance the system's performance, it becomes impossible to do so without eliminating the system conflict.

From a TRIZ standpoint, to make a good invention means to resolve a system conflict without a compromise.

The emergence of system conflicts is deeply rooted in the hierarchical nature of technological systems. This hierarchical structure entirely determines a strategy of technological system development or of inventive problem solving. Any change in a system's structure, even the most needed one, may reverberate with consecutive changes in adjacent systems and subsystems. As a rule, these induced changes are harmful and frequently negate the advantages of the improvement.

Example 2.4
Consider a confluence of system conflicts associated with the development of a highly fuel-efficient car. An obvious direction for improving the fuel efficiency is weight reduction. Experts agree that substantial improvements in this area will be possible only if the steel in car bodies is replaced with lightweight materials. A car with a lighter

Table 2.3. *System conflicts associated with the use of aluminum and plastics in cars*

Aluminum	
Pros	Cons
45%–70% lighter than steel.	Five times more expensive than steel (per unit of weight).
Can be shaped using many conventional steel forming technologies.	Only about one-third the stiffness of steel.
Stiffness can be enhanced by making the panels thicker.	This would negate the weight advantage and increase the costs.

Plastics	
Pros	Cons
Much lighter than steel and aluminum.	Stiffness of ordinary plastics is only $1/60$–$1/30$ that of steel; for reinforced plastics it is $1/15$–$1/5$ that of steel.
Can be formed into a much wider variety of shapes than steel and aluminum.	Up to 10 times more expensive than steel (per unit of weight).
Require 30% fewer parts than metal bodies.	The state of the art assembly processes for plastic structures cannot provide for tight tolerances.

body requires a lighter engine and a less massive suspension. These secondary weight savings can roughly double the benefits: for every 10 pounds saved by reducing the weight of the car body, another 10 pounds can be saved by downsizing other parts of the car (Field & Clark, 1997). The primary material candidates for use in an ultra-light car are aluminum and various plastics. Table 2.3 shows some system conflicts that the automotive industry must overcome before aluminum and plastics become materials of choice for car bodies.

2.5.1   System conflict diagrams

System conflicts can be conveniently represented by graphs indicating useful and harmful actions. These graphs are called *system conflict diagrams*. The system components participating in system conflicts are called *conflicting components*. The most typical system conflict diagrams are shown in Table 2.4.

Every so often, a component may participate in two system conflicts that create a *chain system conflict*, such as the one shown in Fig. 2.8. As a rule, the system conflicts in the chain are time-dependent; i.e., the second system conflict is caused by the first (respectively, the right and left pair of conflicting actions in Fig. 2.8). Chain system

Table 2.4. *Some typical system conflict diagrams (various combinations of these diagrams are also possible)*

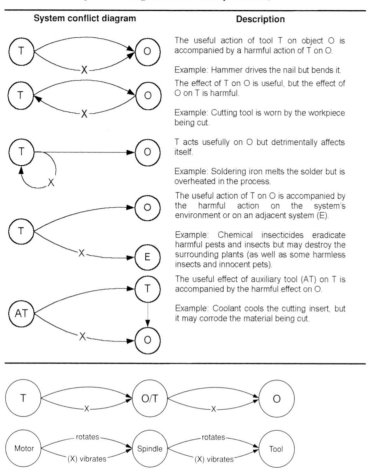

System conflict diagram	Description
	The useful action of tool T on object O is accompanied by a harmful action of T on O.
	Example: Hammer drives the nail but bends it.
	The effect of T on O is useful, but the effect of O on T is harmful.
	Example: Cutting tool is worn by the workpiece being cut.
	T acts usefully on O but detrimentally affects itself.
	Example: Soldering iron melts the solder but is overheated in the process.
	The useful action of T on O is accompanied by the harmful action on the system's environment or on an adjacent system (E).
	Example: Chemical insecticides eradicate harmful pests and insects but may destroy the surrounding plants (as well as some harmless insects and innocent pets).
	The useful effect of auxiliary tool (AT) on T is accompanied by the harmful effect on O.
	Example: Coolant cools the cutting insert, but it may corrode the material being cut.

Fig. 2.8. Chain system conflict.

conflicts are usually decomposed into constituent system conflicts, and each is addressed separately.

2.6 Resolving system conflicts

Consider any typical system conflict diagram from Table 2.4. *To resolve a system conflict means to eliminate the harmful action while retaining the useful action.* Notice that the elimination of the harmful action is equivalent to the introduction of a useful one (e.g., prevention of the overheating of the solder iron is usually equivalent to maintaining its correct temperature).

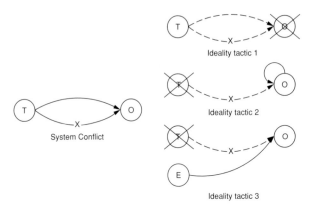

Fig. 2.9. Ideality tactics for resolving system conflicts.

Three generic approaches are used for resolving system conflicts.
(1) Elimination of either the tool or the object (using the ideality concept).
(2) Changing the conflicting components (i.e., the tool or the object) such that the harmful action disappears.
(3) Introducing a special component (tool) intended to eliminate or neutralize the harmful action.
Next, we consider all of these approaches.

2.6.1   Resolving system conflicts: using the ideality concept

This approach follows directly from the concept of the ideal technological system. Removing either the tool or the object automatically eliminates the harmful action. Since the useful action must still be performed, some material bodies should be responsible for this performance. This can be realized by the following *ideality tactics* (Fig. 2.9).

Ideality tactic 1
 The object is eliminated, which causes the elimination of the tool and of all the useful and harmful actions.

Ideality tactic 2
 The tool is eliminated, and the object (or one of its parts) itself performs the useful action.

Ideality tactic 3
 The tool is eliminated, and its useful action is delegated to the environment or to another tool.

Application of the concept of ideality frequently requires the recognition and use of *system resources*, many of which may not initially be obvious. Many inventions developed by TRIZ practitioners come from the recognition and use of previously unnoticed resources.

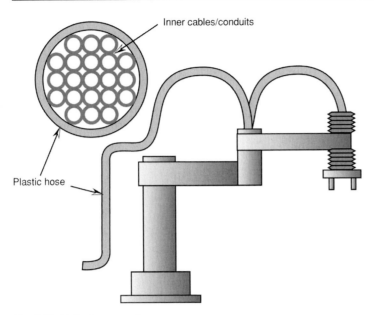

Fig. 2.10. Robotic test station.

It is also important to understand that the realization of "ideality" in real life comes with a price. Application of these ideality tactics usually requires minor changes in other system components (Problem 2.2 illustrates this when ideality tactic 1 is used). Similarly, reassigning the useful action to the available resources – either to the object or to the environment (ideality tactics 2 and 3, respectively) – will, most likely, cause changes in the object and environment, or in their connections with other system components.

Problem 2.2

In a robotic test station for computer components, the end effector of the robot (Fig. 2.10) performs complex manipulations (clamping, handling, and inserting) with very delicate parts in a clean room. It is energized by two vacuum lines and four compressed air lines and has several sensors transmitting signals via electrical cables. All vacuum, compressed air, and electrical communications are channeled to the distribution/control box at the robot's base by a corrugated plastic hose encasing all the tubes and wires. The purpose of this hose is to prevent the contamination of the clean room by containing fine particles generated by any rubbing between the tubes and wires. The hose has shown a tendency to rupture in service. The rupture was due to fatigue associated with large amplitude, high-speed link motions resulting in an excessive twisting of the hose. This allowed the wear particles to escape into the environment and caused excessive contamination due to the rubbing between the ruptured surfaces themselves.

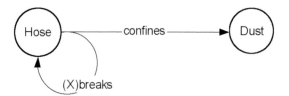

Fig. 2.11. System conflict: the hose confines the dust, but its own movements cause it to break.

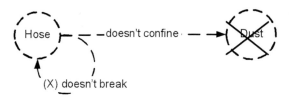

Fig. 2.12. Absent dust renders the hose unnecessary.

The test station manufacturer incurred large warranty costs (sending technicians to the customer locations to replace ruptured hoses every 2–3 months). The company was trying to improve the situation by solving the problem: *How to prevent the breakage of the hose?* Use of more durable plastics for the hose was more expensive, and did not significantly expand its useful life. Creating a negative air pressure in the hose was suggested, thus preventing the dust particles from escaping but that, also, would have increased costs.

The authors were asked to find a way to extend the service life of the hose and reduce the costs.

Analysis
The system conflict for this situation is shown in Fig. 2.11. The only hose that cannot be broken or damaged and does not need any service is an absent one. The ideal hose should not exist. This is possible if the dust particles are not generated in the first place (Fig. 2.12).

Solution
To prevent dust generation, i.e., rubbing between the inner conduits and wires, they must be separated. A simple way to accomplish this separation is by using elastic supports (Fig. 2.13).

Problem 2.3
A "rough idle" problem was harming the sales of, and creating high warranty costs for, a compact car model in the early 1990s.

Four-cylinder engines in compact cars generate unbalanced dynamic forces (second order). At the engine's idle speed (450–900 r.p.m.), the second-order frequency is very

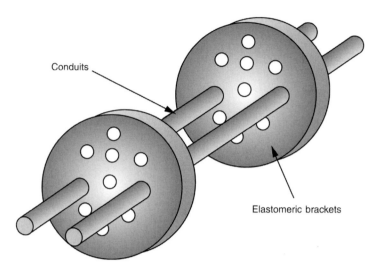

Conduits

Elastomeric brackets

Fig. 2.13. Elastomeric dividers separate the conduits and cables, thus preventing dust generation.

close to the natural frequency of the steering column, resulting in a violent shaking of the steering wheel. The unbalanced forces can be reduced by a special balancer, but at higher production costs. These "rough idle" vibrations became especially objectionable after adding a driver air bag. This addition shifted down the natural frequency of the steering column to resonate with the second-order forces at idle r.p.m.

A rather costly reinforcement of the steering column alleviated the shake, but its intensity was still unacceptable. A dynamic vibration absorber (DVA) with a 1-pound lead inertia mass – was attached to the steering column, but the shake was still much more violent than in other compact cars on the market. No space was available for a larger inertia mass. Increasing the idle rotational speed was ruled out, since that would have impaired the fuel efficiency.

The main causes of the problem were identified as high engine power at idle due to inefficiencies of the engine itself, as well as of the air conditioner and alternator, and a high drag torque of the automatic transmission.

A task force composed of senior engineers from engine, transmission and electrical divisions worked for several months and mapped out long-range projects (maxi-problems) to solve the identified problems. Millions of dollars in warranty costs, however, kept mounting.

A logical conclusion was to eliminate, or at least alleviate, these factors. All of the divisions involved launched projects aimed at reducing energy losses in the identified critical units and at stiffening the body. Certainly, these projects were needed, not only to cure the "rough idle" problem, but for a general quality improvement of the compact car and its successors. These major projects (maxi-problems), however, were not promising any results before the expiration of the life cycle of this model.

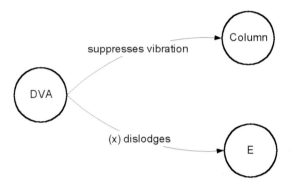

Fig. 2.14. System conflict: DVA suppresses the shake but displaces the surrounding components.

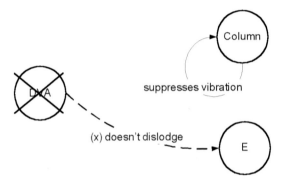

Fig. 2.15. Ideal DVA is absent.

At about the same time as the decision to go ahead with these projects was made, one of the authors was invited to join the task force. The TRIZ analysis was conducted in line with the mini-problem approach and ideality tactic 2.

Analysis
Since an immediate solution was required, the only quick, practical approach was to suppress the shake with an effective DVA. The system conflict: a DVA with a large inertia mass could adequately abate the shake of the steering column, but it was unacceptable (Fig. 2.14).

The ideal inertia mass is absent, but its function is fully performed. This called for removing the existing lead block and assigning its function to a resource component elsewhere in the car, preferably to the steering column itself (Fig. 2.15).

Solution
In a complex dynamic system, an appropriately tuned damper may reduce the vibration intensity of a component not directly connected to, or even located remotely to it.

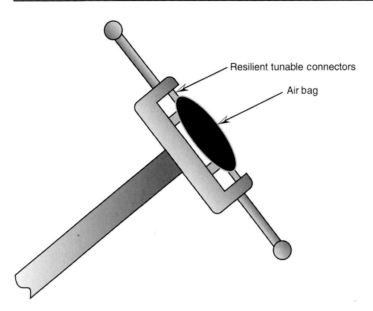

Resilient tunable connectors

Air bag

Fig. 2.16. Airbag, attached to the steering column via tunable connectors, acts as an effective DVA.

The important parameters determining performance characteristics of a DVA are its tuning and the "mass ratio" between its mass and the effective mass of the element whose vibration is reduced. After performing the TRIZ analysis, the car structure was evaluated and massive structural components that could be used as an inertia mass were identified (air bag, spare tire, battery, etc.).

Tests showed that the air bag was the most effective component. To use the air bag as an inertia mass for a DVA, it was separated from the steering column by flexible tunable connectors (Fig. 2.16). This reduced the effective mass of the column from 7 lbs down to 3.3 lbs and resulted in a mass ratio of about 1.0, as compared with the mass ratio of DVA with the lead inertia weight of 0.14. Consequently, the shake amplitudes of the steering wheel were reduced six to seven times below those of the (then) best-in-class Toyota Corolla.

Extensive testing confirmed that this "additional burden" did not impair the primary function of the air bag (US Patent No. 6,164,689).

Problem 2.4
Self-unloading barges are often used in dam construction (Fig. 2.17). The barge carries the bulk cargo (e.g., stones, pebbles, etc.) and is hauled by a tugboat to the place where the cargo should be unloaded (by turning the barge upside down). The heavy keel and the buoyancy force generate a moment, returning the empty barge to its upright position. The heavier the keel is, the faster the upturn, however the heavy

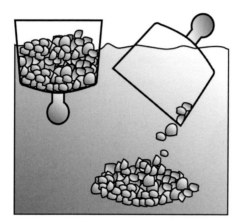

Fig. 2.17. Self-unloading barge (not to scale).

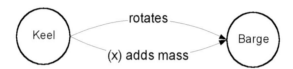

Fig. 2.18. Heavy keel is good for upturning the barge, but it increases its weight.

keel reduces the weight-carrying capacity of the barge, which is an undesirable effect.

Analysis

The system conflict: the heavy keel creates the moment that upturns the barge, but it also adds weight (Fig. 2.18). This conflict can be resolved if the function of the keel is assigned to the environment (ideality tactics 3).

Solution

The hollow tank-keel is filled with water. In the upright position, such a keel does not contribute an additional weight, but being lifted in the air, it helps to turn the barge upright (Fig. 2.19).

2.6.2   The ideality tactics and auxiliary tools

All of the above problems share one common element: each of them was resolved by eliminating an auxiliary tool (the hose in Problem 2.2, the lead block in Problem 2.3, and most of the heavy keel – in Problem 2.4).

Let's introduce a rule.

A system conflict involving an auxiliary tool can be resolved by eliminating this tool and reassigning its action to one of the major components of the system, or to the environment.

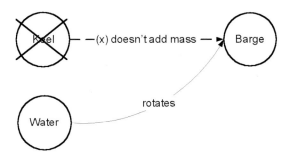

Fig. 2.19. "Water keel" flips the barge without adding the weight.

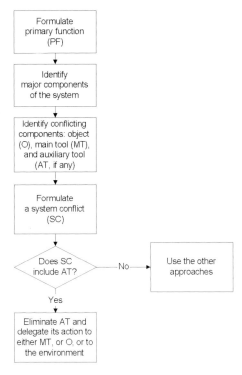

Fig. 2.20. Elimination of an auxiliary tool.

Since most technological systems have at least one auxiliary tool, this rule is used very often. The first step in applying this rule is the formulation of the primary function of the system. Then, all major components should be listed and conflicting components identified. If one of the conflicting components is an auxiliary tool, it should be removed and its action should be delegated as per the rule (Fig. 2.20).

If a system conflict diagram includes more than one auxiliary tool, it is recommended to begin the "clipping" process with correcting and measuring auxiliary tools, then move

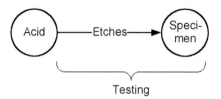

Fig. 2.21. Primary function of the setup.

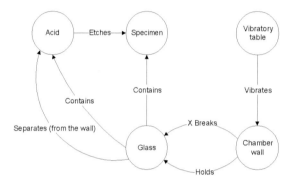

Fig. 2.22. Conflicting interactions between components.

to enhancing tools and, finally, to enabling ones (see Section 2.2). If, for some sound reason, an auxiliary tool cannot be eliminated, then it should be modified.

Consider Problem 2.1

The primary function of this setup is the testing of alloys (specimens) in the presence of vibration (Fig. 2.21). The major components are: the specimen(s), acid, glass, chamber walls, and a vibratory table (not shown).

    The conflicting components are: the specimen (object), acid (main tool), chamber and vibratory table (enabling auxiliary tools) and glass (correcting auxiliary tool). The system conflict diagram is shown in Fig. 2.22 (vibrations propagating from the chamber wall onto other components are not shown to simplify the diagram).

    Because the glass is a correcting auxiliary tool, it should be eliminated first. In addition to performing its original action – separating the acid from the chamber walls – the glass (together with the chamber wall), in effect, contains the acid and specimens. The latter action also disappears with the elimination of the glass.

Solution

The acid is poured into an appropriately shaped specimen that becomes a "chamber" (Fig. 2.23). The required conditions for carrying out the tests, such as pressure and temperature, are created outside the specimen chamber.

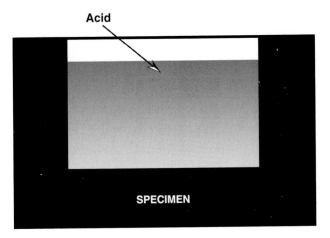

Acid

SPECIMEN

Fig. 2.23. Ideal solution.

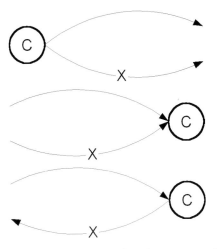

Fig. 2.24. Possible interfaces between a conflicting component and useful and harmful actions.

2.6.3   Resolving system conflicts: changing the conflicting components

Physical contradictions

A system conflict takes place *between two or more conflicting components*, as is seen in Table 2.4. When trying to overcome a system conflict, *one* of these conflicting components can often be changed or eliminated. The component C in Fig. 2.24, has to perform one of the three possible pairs of actions.

(1) Providing the useful action *without* generating the harmful one.

(2) Receiving the useful action *and* protecting itself from the harmful one.

(3) Receiving the useful action *and* not generating the harmful one.

Often, to perform one of these pairs of actions, a conflicting component must be in opposite physical states, or it must posses mutually exclusive properties, e.g., it has to be *hot* and *cold*, *electroconductive* and *electroinsulative*, *heavy* and *light*, *present* and *absent*, etc. For example, a heavy (bulky) DVA inertia mass (Problem 2.2) must be present to suppress the shake, and it must be absent so as not to affect the packaging of other components. Or, the keel in Problem 2.3 must have a certain weight to flip the barge *and* it must be weightless so as not to increase the weight of the barge.

A situation in which the same component ought to be in mutually exclusive physical states is called a physical contradiction.

A physical contradiction is formulated by the pattern: "*To perform action A_1, the component (or its part) must have property P, but to perform (prevent, neutralize) action A_2, this component (or its part) must have an opposite property –P.*"

Opposing demands formulated for the whole component are called a *macro physical contradiction,* while opposing demands formulated for the component's constitutive elements (particles or segments) are called a *micro physical contradiction.* For example, the statement "To perform action A_1 the component must be hot, and to prevent action A_2 the component must be cold," is a macro physical contradiction. However, for the component to be hot its particles have to move quickly; particles of a cold component would move slowly. Thus, the first physical contradiction can be reformulated: "To perform action A_1 the particles of the component must move quickly, and to prevent action A_2 these particles must move slowly."

Conventional design philosophy implies an inevitability of compromises: if a component has to be both snow white and pitch black, it usually ends up being a shade of gray. Contrary to this approach, TRIZ offers specific ways out of the impasse. Since the same component cannot possess mutually exclusive properties *in the same point of space* or *at the same time,* three generic *separation principles* can be used.

Separation of opposite properties in time:
at one time a component has property P, and at another time it has an opposite property –P.
Separation of opposite properties in space:
one part of a component has property P, while another part has an opposite property –P.
Separation of opposite properties between the whole and its parts:
a system has property P, while its components have an opposite property –P.

Often, actions are performed most effectively when *each action is carried out by a designated physical entity*: action A_1 by one entity, action A_2 by another entity. These physical entities come in various forms and shapes: as the whole component, or the surface of this component, or a group of molecules or atoms, or a solo elementary particle, etc. Often, one of these entities is already given, so the resolution of a physical contradiction is reduced to generating (or introducing) another entity.

To solve a problem that contains a physical contradiction, one should do the following.

(1) Identify the conflicting component associated with both useful and harmful actions (i.e., with a system conflict).
(2) Formulate a pair of mutually exclusive demands to the physical state (properties) of this component.
(3) Attempt the separation principles – individually or in combination – to satisfy these demands.

Separation of opposite properties in time

This separation principle is often based on the transition to flexible structures (in the sense of the law of increasing dynamism [flexibility], mentioned in Chapter 1 and described in Chapter 5). The more flexible the component, the more fully its contradictory properties can be separated.

The use of segmented linkages, flexible materials, smart substances, nonlinear components and electromagnetic fields allows for the temporal separation of contradictory demands.

Handheld keyboards should satisfy two conflicting demands. To be ergonomical, the keyboard should have sufficiently large keys or a keypad; however, the use of a full size keyboard defeats the main purpose of the portable electronic device.

The first step toward overcoming this conflict was the introduction of foldable keyboards that occupy minimum space when not in use (Fig. 2.25). Next, Logitech Inc. brought in a fabric-based keyboard that also serves as a carrying case for a handheld. The next innovation, according to the line of transition to fields, should be a keyboard based on electromagnetic fields. Interestingly enough, one such device was recently developed by the Israeli company VKB, Ltd. This is a keyboard made entirely of . . . light. An image of a full QWERTY keyboard is projected onto a flat surface in front of the user. VKB has developed an efficient method for projecting an optical image of a keyboard onto a surface. A proprietary detection method allows for the accurate and reliable detection of user interaction, such as typing or cursor control functions (e.g. mouse or touch-pad controls).

For most mechanical systems, there are other powerful techniques that help realize separation in time:

• *Phase transformations* (e.g., transitions from a solid to a liquid, from a liquid to a gas, etc.).

This technique is used in deactivating retail security tags. A typical security tag is a resonant circuit (coil) imprinted on the surface of a substrate (Fig. 2.26). The coil's parameters are tuned to a predetermined radio frequency detection signal transmitted by an alarm security system. When merchandise with an attached security tag passes the checkpoint area, the tag begins to resonate from the transmitted energy, resulting in the activation of audible and/or visible alarms. Once the merchandise has been purchased, however, it is necessary either to detach or deactivate the security tag so that the merchandise can be removed from the store.

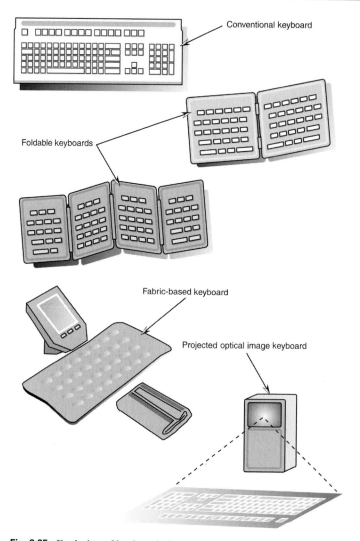

Fig. 2.25. Evolution of keyboards for handhelds.

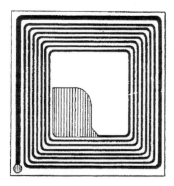

Fig. 2.26. Security tag.

The physical contradiction – the tag must be present and it must be absent – is resolved with the help of a fusible link that melts when the tag is exposed to energy above a certain level. Thus, upon purchase of the tagged merchandise, the coil is deactivated by exposing the security tag to electromagnetic energy sufficient to destroy the fusible link.

• *Decomposition of substances*: at a certain moment, the component material (substance) is broken down into its elements (molecules, atoms, etc.) which perform one of the opposing actions.

Problem 2.5

Radio antennas used in some military applications (e.g., by the Special Forces operating deep behind enemy lines) can be detected by enemy radars. How can one make these antennas transparent to incoming radar signals?

Analysis

Radar signals (i.e., radio waves) bounce off conventional antennas because the latter are metallic, that is electro-conductive. Electrical conductivity is necessary because it enables the antennas to emit and receive "desirable" radio waves. Non-conductive materials are transparent to radar signals but they cannot emit/absorb "desirable" radio waves. Thus, to function, an antenna must be conductive, but to be stealthy it must be non-conductive. This physical contradiction should be resolved in time: the antenna is normally non-conductive, except for short periods when it is in transmission/reception mode.

Electrical conductivity of metals is due to free electrons moving inside their crystal lattice. Understanding this condition helps reformulate the initial *macro physical contradiction* into a *micro physical contradiction*: "Free electrons must be present in the antenna, and they must be absent." Separating these demands in time means that electrons are generated only when the antenna is on; when it is off, the electrons are bound to the atoms of the antenna's material.

Solution

The novel antenna is a plastic or glass tube containing a low-pressure gas (basically, it is a fluorescent light bulb). When the gas is ionized (i.e., its molecules are broken down into ions and electrons in the presence of high voltage), it becomes conductive, allowing radio signals to be transmitted or received. When the gas is not ionized (the voltage is off), the antenna ceases to exist as a conducting medium (http:\\www.marklandtech.com\gasplasma.html; Moisan *et al.*, 1982).

Separation of opposite properties in space

This separation principle entails transition from mono- to bi-systems, in which each sub-system (component) possesses one of the opposing properties (and performs one of the conflicting actions).

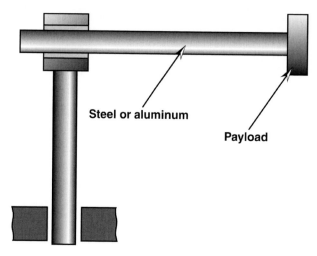

Fig. 2.27. Revolving manipulator link.

Problem 2.6

A revolving link of a pick-and-place manipulator (Fig. 2.27) has to satisfy mutually exclusive requirements. It should be light (for a given shape) to reduce the size (and weight) of driving motors and/or to allow for larger payloads. Light materials, however, usually have reduced stiffness (Young's modulus). On the other hand, the link should be rigid, to reduce its deflections, as well as settling time in start/stop periods. Rigid materials, though, are usually heavy.

Analysis

The physical contradiction is given: the link must be light, but it must be heavy (rigid). A simple engineering analysis in Fig. 2.28 shows that the intensity of inertia forces caused by the acceleration/deceleration of the link is greatest at its distal (farthest from the joint) parts, while the bending moment intensity is maximal at its proximal (closest to the joint) parts. Accordingly, the link's deflection is determined by its root segment for angular deformations (the highest bending moment; the longest projection length), whereas the link inertia is determined mainly by the overhang segment. Thus, the contradictory requirements are separated in space.

Solution

The link is composed of two sections (Fig. 2.28): the root section, which does not significantly influence the effective mass of the link but determines its stiffness, is made from a high Young's modulus material (e.g., steel), while the overhang section, which does not noticeably contribute to stiffness but determines the effective mass, is light, e.g., made from aluminum (Rivin, 1988).

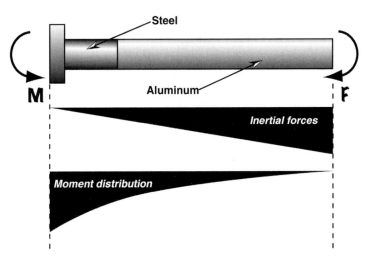

Fig. 2.28. Composite revolving link and distribution of the bending moment and inertia forces along the link.

Note that the same initial situation can be associated with different physical contradictions. For example, for Problem 2.6, one can formulate another physical contradiction: the link must be long, to deliver the payload to the required place, and it must be short, to reduce the deflections and the settling time. These contradictory requirements can be separated in time. A telescopic arm resolves the contradiction. Its sliding links are pulled in during rotation periods and are extended between the rotations.

Separation of opposite properties between the whole and its parts

This separation principle proves to be particularly helpful when resolving such physical contradictions as rigid vs. compliant, hard vs. soft, etc. In some cases, solutions to such contradictions employ flexible and loose materials: films, fibers, pellets, granules, powder, etc.

Problem 2.7

A conventional bench vise is designed to clamp parts of regular shapes (Fig. 2.29). To clamp irregularly shaped parts, special jaws have to be installed. The fabrication of such jaws is usually costly.

Analysis

Physical contradiction: the jaws must be rigid to clamp the part, and they must be flexible to accommodate themselves to the part's geometry. The opposite properties can be separated between the system and its parts: the jaws are replaced with a fragmented (and, therefore, flexible) medium, whose individual components are rigid.

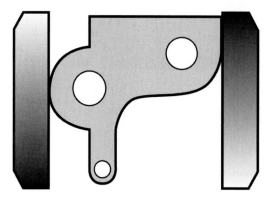

Fig. 2.29. Conventional vise and an irregularly shaped part.

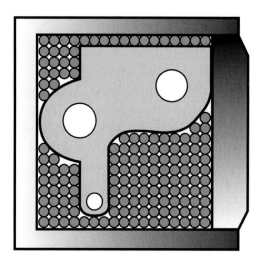

Fig. 2.30. Adaptive vise.

Solution

One of the embodiments of this approach involves small balls (Fig. 2.30). These elements can accommodate virtually any contour and, as an additional benefit, distribute the clamping force uniformly.

In another embodiment of the same approach, small particles are packed in a pouch. After the part is aligned, the air is sucked from the pouch, and the particles are cemented into a rigid body of the required shape by friction forces due to the atmospheric pressure.

Other vise designs that realize this approach use movable pins or fingers (see, for example, US Patents 5,145,157 and 5,746,423).

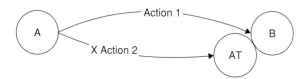

Fig. 2.31. Blocking the harmful action.

Fig. 2.32. Sealing ampoules with drugs.

2.6.4   Resolving system conflicts: introducing a new tool for eliminating or neutralizing the harmful action

A new (auxiliary) tool usually performs three actions: *blocking the harmful action, counteracting the harmful action, drawing the harmful action away from the affected component.* These actions are often used in conjunction with the separation principles.
• *Blocking the harmful action* (Fig. 2.31).

Problem 2.8
Ampoules containing drugs are flame-sealed, as shown in Fig. 2.32. The flame is difficult to control; intense flame may overheat the drug, while a small flame may be insufficient to melt the glass.

Analysis
It is obvious that these contradictory demands should be separated in space – along the ampoule's height: intense flame is concentrated only around the ampoule's neck, and is blocked from the rest of the ampoule.

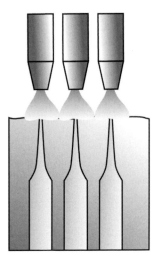

Fig. 2.33. Water cuts the flame off from the drugs.

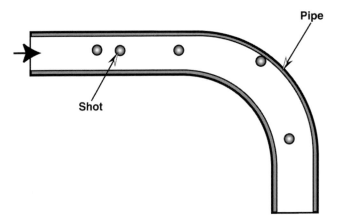

Fig. 2.34. Shots destroy the pipe.

Solution
The problem was solved by immersing the ampoules in water (Fig. 2.33). A very intense flame assures that the ampoules are sealed, but does not affect the drug.

Often, the blocking auxiliary tool can be made from already available resources, sometimes from the conflicting components themselves.

Problem 2.9
Curved sections of pipes in steel shot-blasting machines wear very rapidly (Fig. 2.34). Various protective coatings are minimally effective, since none can hold out against the shot bombardment long enough.

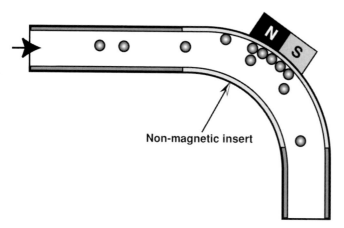

Fig. 2.35. Self-protecting system.

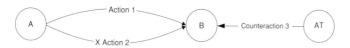

Fig. 2.36. Counteracting the harmful action.

Analysis
Since "foreign" protective coatings do not work, it is logical to try internal resources.

Solution
The shot particles themselves serve as an intermediary. A magnet is placed outside the wear zone. It attracts some flying shot particles that form the protective layer (Fig. 2.35).
• *Counteracting the harmful action* (Fig. 2.36).

Problem 2.10
Punching holes in thin-walled tubes (Fig. 2.37) may result in the collapse of the tube walls and, consequently, in forming poorly defined holes. Supporting the tube walls with a solid could prevent their collapse; in a tube with many bends, however, it may be difficult to access the tube interior so as to provide support. An obvious way out – punching the holes before bending a tube – is not always suitable, since bending can distort the holes.

Analysis
To support the walls, a back-up has to be solid; it has to be non-solid (fluid) in order to be easily positioned inside the tube. (An equivalent, but shorter, formulation: a back-up

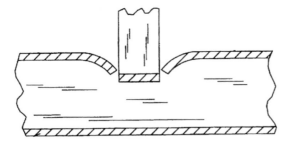

Fig. 2.37. Collapsing of tube walls.

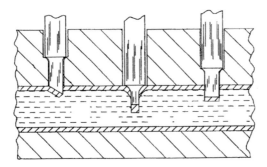

Fig. 2.38. Ice precludes collapse of the tube walls (US Patent 5,974,846).

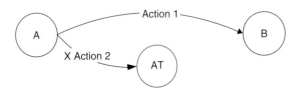

Fig. 2.39. Drawing the harmful action away from the affected component.

has to be present, and it has to be absent). Since these contradictory demands have to be satisfied in the same place, they should be separated in time.

Solution
A process has been developed by Greenville Tool and Die Co: prior to piercing, a tube is filled with a liquid (e.g., water), which is then frozen. The solidified liquid provides support to the tube wall during piercing. After the piercing, the solidified liquid is melted and drained from the tube (Fig. 2.38).
• *Drawing the harmful action away from the affected component* (Fig. 2.39).

Problem 2.11
How to prevent the bursting of water pipes exposed to sub-zero temperatures? Heating a pipe or reinforcing its walls would add excessive complexity and cost.

Fig. 2.40. Insert for preventing rupture of pipes (US Patent 6,338,364).

Analysis
The problem can be resolved if the compressive force of the expanding freezing water is redirected toward another component.

Solution
According to US Patent 6,338,364, a round, flexible plastic insert is placed inside a water pipe (Fig. 2.40). Guides secure an axial position of the insert within the pipe. The insert maintains its original shape when the water is in a fluid state. As the water freezes, the insert is compressed, thus accommodating the volumetric expansion of the freezing water, which prevents the structural failure of the pipe. When the ice melts, the insert flexes back to its original shape.

2.7   Summary

Resolving system conflicts is an essential part of both problem-solving and the development of new technological systems. Among the three approaches for resolving system conflicts – the ideality tactics, overcoming physical contradictions and the introduction of a new tool – the first one is the most preferable because it leads to the simplification of the system. Once a system conflict has been formulated, this approach should be considered first. If this approach is not applicable, then the other two approaches should be attempted. In this case, one should strive to keep the system as simple as possible – first by using readily available resources (Fig. 2.41).

2.8   Exercises

These exercises (as well as exercises in the following chapters) are offered to improve your problem analysis and concept development skills. Try not to guess the right answer (even if it appears to be obvious), but look for it using the concepts and tools of TRIZ.

2.1 Formulate primary functions performed by:
- Shaving razor
- Toothbrush

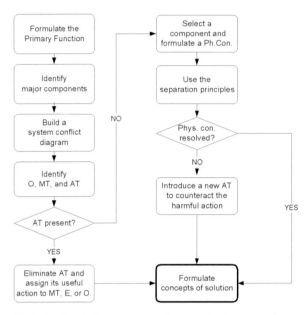

Fig. 2.41. Resolving system conflicts: putting it all together.

- Mailbox
- Computer keyboard
- Flashlight
- Ski.

2.2 Rank functions performed by different components of:
- Shaving razor
- Toothbrush
- Mailbox
- Computer keyboard
- Flashlight
- Ski.

2.3 Using the ideality tactics, develop alternative conceptual solutions to Problem 2.1.

2.4 Heat produced by IC chips travels through the leads and heat sinks – small rods made of the solder material (Fig. 2.42). The heat sinks are fabricated before the IC chips are soldered to the printed circuit board and, therefore, they also evacuate heat from the soldering zone during the melting of the solder. The loss of heat during soldering may have a detrimental effect on the mechanical strength of the solder joint. Thus, the heat sinks assist in transferring the heat flow from the IC chip when it functions, but they reduce the solder temperature at the time of board fabrication (due to certain production constraints, it is very undesirable to fabricate the heat sinks after the soldering operation). Use the ideality tactics to resolve this situation.

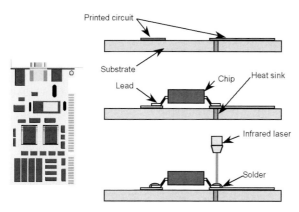

Fig. 2.42. Soldering an IC to a PCB.

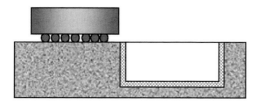

Fig. 2.43. The blast furnace needs to be lowered without special equipment.

2.5 Formulate and resolve a physical contradiction in the following problem.

To plate a metal part with another metal, the part is placed in a bath filled with a salt solution of the coating metal. The solution is heated and, due to thermal decomposition, the metal from the solution precipitates onto the part's surface. The speed of coating is temperature-dependent, however: at high temperatures the solution decomposes so fast that up to 75% of the salt is wasted, lodging on the bottom and on the walls of the bath. Lowering the temperature leads to an unacceptable reduction of the production rate.

2.6 Auguste Piccard, a famous Swiss inventor, was developing a bathyscaph – a submarine vessel intended to explore great ocean depths. In 1960, the bathyscaph *Trieste*, designed by Piccard and operated by his son Jacques and Lt. Donald Walsh of the US Navy, descended 10 916 m to the bottom of the Marianna Trench, the deepest known point in the Earth's oceans.

Before Piccard, deep-water descents had been made with a bathysphere, a thick-walled metal chamber suspended by a steel cable from a surface ship. In the case of a cable rupture, the bathysphere would be impossible to lift from the bottom. Piccard decided to design a vessel capable of autonomous operation. For this vessel to surface, it had to be lighter than water. That meant it that it had to have a large volume and thin walls. In order to prevent it being squashed by the tremendous water pressure at great depths, however, the vessel had to have a small surface area and thick walls.

"Reinvent" the solution discovered by Piccard.

2.7 The 230 000-lbs base of a blast furnace is moved next to a foundation well (Fig. 2.43). How should the base be lowered into the well? Of course, special cranes can be used, but this would significantly increase the complexity and cost of the operation, and may not be reliable. What should be done?

2.8 Magnetic resonance imaging (MRI) scanners use strong magnetic fields and powerful radio wave pulses. MRI cannot be used on patients who have pacemakers. These pacemakers are usually enclosed in a metallic cover that shields them from outside electromagnetic radiation. The problem is that radio waves, generated by an MRI scanner, can induce eddy currents in the cover that are large enough to heat it to over 70°C (~160°F). This is more than hot enough to damage surrounding tissues. Use the template in Fig. 2.41 to resolve this situation.

3 Basics of the substance–field analysis

Physical phenomena are the basis for all technological systems. To perform a function, certain physical phenomena should be appropriately arranged in space and time. For example, to move the bristles of an electric toothbrush, the battery supplies electric energy to the motor, which then converts it into the rotational movement of the rotor which, in turn, actuates a transmission link, setting the brush head in motion.

To improve an existing function, or to introduce a new one, means to make a transition from a particular system's physical structure and/or physics to another, more effective structure and/or physics. As an example of such a transition, consider the external combustion (steam) engine and the internal combustion engine. In the former, the piston is moved by high-pressure steam produced outside of the engine cylinder. In the latter, the fuel burned inside the cylinder moves the piston. In addition to different physical structures, these engines employ different physical phenomena for generating motion.

This chapter describes a modeling approach used in TRIZ for the analysis and synthesis of physical structures and processes in technological systems.

3.1 Minimal technological system. Substance–field models

Any interaction between a tool and an object is accompanied by the generation, absorption, or transformation of energy. Thus, the object, the tool, and the energy of their interaction are necessary and sufficient to build a model of a *minimal technological system* performing only one function (Fig. 3.1). *An effectively performed function requires the presence of, and interaction among, these three elements.*

The triad "object–tool–energy" is described in TRIZ in terms of *substances* and *fields*.

The term "substance" has a rather broad meaning. Frequently, a substance is a technological system of varying degrees of complexity, for instance, a nail, a keyboard, a ship and so on.

Table 3.1. *Fields used in TRIZ*

Fields				
Mechanical	Thermal	Electrical	Magnetic	Chemical
• Friction force • Centrifugal force • Vibration • Ultrasonic • Stress • Pressure gradient	• Infrared radiation • Temperature gradient	• Monopole • Dipole • Line charge • Line dipole • Traveling • Polarized	• Permanent • Straight-wire field • Solenoidal field • Magnetic dipole	• Chemical-potential gradient • Surface energy

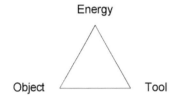

Fig. 3.1. Elements of a minimal technological system.

The term "field" refers to the energy needed for the interaction between two substances. Modern physics recognizes four fundamental fields (forces).

(1) Gravitational field (acts between any two or more pieces of matter).

(2) Electromagnetic field (causes electric and magnetic effects, binds atoms and molecules together).

(3) Weak nuclear field (causes radioactive decay).

(4) Strong nuclear field (responsible for holding the nuclei of atoms together).

However, this classification is not always sufficient when solving conceptual design problems. The meaning of a field in TRIZ is totally different, and allows one to conveniently distinguish between various interactions. TRIZ operates with *engineering fields*, such as *mechanical, thermal, electric, magnetic*, and *chemical*. These fields – *which are always generated by substances* – manifest themselves through many groups of physical and chemical phenomena, some of which are listed in Table 3.1.

An object substance is usually depicted as S_1, a tool substance as S_2, and a field as **F**. The substance–field structure, composed of two substances, S_1 and S_2, and an interaction field, **F**, is called a *sufield* (*substance + field*, Fig. 3.2).

The notations and lines used in the sufield analysis are shown in Table 3.2.

A substance can generate a field (Fig. 3.3a), or a field can be applied to a substance (Fig. 3.3b) or a substance can transform one field into another (Fig. 3.3c).

Table 3.2. *Sufield notations*

Sufield notation	Description
Δ	Sufield
_____	Unspecified action
⟶	Specified action
–·–·–·–►	Inadequate action
⟶×⟶	Harmful action
⟹	Transition from an initial sufield structure to the desirable one
→ **F**	Field generated by a substance
F→	Field applied to a substance
F'	Modified field
S'	Modified substance
(S₁ S₂)	Attached substances (e.g., a bi-metal strip)

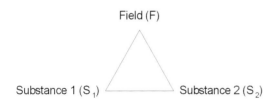

Field (F)

Substance 1 (S$_1$)   Substance 2 (S$_2$)

Fig. 3.2. Simplest complete substance–field structure.

Substances in sufields interact with each other via fields. These interactions have to be shown for a precise depiction of a sufield (Fig. 3.4).

This formula reads as follows: Field F_1 acts on substance S_2 that transforms it into field F_2 that acts on substance S_1. Field F_2 can be of the same or of a different nature as F_1.

For example, a system "hammer driving nail into the wall" can be represented by a sufield (Fig. 3.5), in which F_{mech} is a mechanical energy applied to the hammer, S_2, which directs it to the nail, S_1.

For the sake of simplicity, in the following text, fields acting between two substances will be shown only in cases when F_2 differs from F_1.

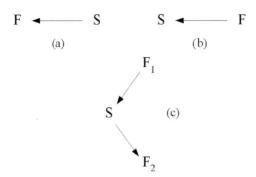

Fig. 3.3. Possible interactions between fields and substances.

Fig. 3.4. Substances interact via fields.

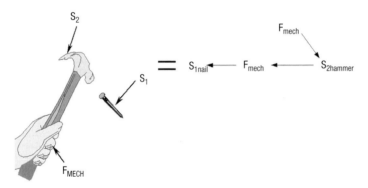

Fig. 3.5. Sufield model of a simple mechanical system.

The most interesting design solutions usually deal with the transformations of an input field into a field of another nature (e.g., a thermal field into a mechanical field; a mechanical field into a magnetic; etc.).

In some cases, it is not easy to recognize the tool, S_2, in a sufield model. This usually happens when S_2 "hides" inside the object, S_1, e.g., it is a part of S_1 performing its action at a level of the object's particles (*micro-level, see* Chapter 5).

Induction heating of a metal part can be modeled as shown in Fig. 3.6: a high-frequency electromagnetic field directly heats the part.

A more thorough analysis of the physical mechanism of induction heating, however, reveals an intermediary structure inside the part – specifically, a collection of free

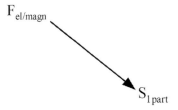

Fig. 3.6. Sufield modeling of the induction heating process with two elements.

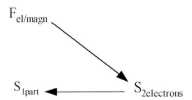

Fig. 3.7. Sufield modeling of the induction heating process with three elements.

electrons. The agitation of these electrons by the electromagnetic field causes the heating of the whole part. This process is modeled as a sufield triad in Fig. 3.7.

The branch of TRIZ studying substance–field structures and their transformations is called *substance–field (sufield) analysis* (Altshuller *et al.*, 1973).

3.2   Modeling physical effects as sufields

The sufield language enables a convenient – from the TRIZ perspective – modeling of physical interactions (physical effects) in technological systems (Altshuller, 1988a). From a sufield analysis standpoint, a physical effect is a certain set of interactions between a substance (or a group of substances) and fields.

The general model of a physical effect is shown in Fig. 3.8. It includes a substance (or a group of two or more substances) that transforms input fields F_{input} into output fields F_{output}. The substance may also change its parameters P (e.g., shape, dimensions, coefficient of friction, magnetic permeability, refraction index, density, position in space, etc.). The model includes control fields $F_{control}$ that can modulate (increase, suppress, turn on/off, etc.) actions of the input fields F_{input}.

The sufield model of the effect of thermal expansion (change in dimensions of a material resulting from a change in temperature) is shown in Fig. 3.9.

Figure 3.10 depicts the sufield model of the effect of *piezoelectricity*, which is the ability of some crystals to generate a voltage under mechanical stress. This effect is used in electric cigarette lighters and sparkers for gas grills: pressing the button compresses a piezoelectric crystal that generates the high voltage, producing the spark that ignites the gas. It is also used in piezoelectric microphones, in load sensors, and in numerous

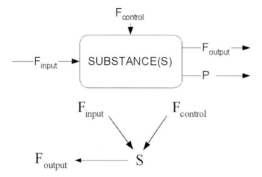

Fig. 3.8. General sufield models of a physical effect.

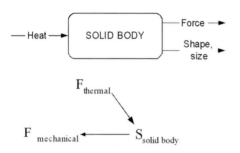

Fig. 3.9. Sufield models of the effect of thermal expansion.

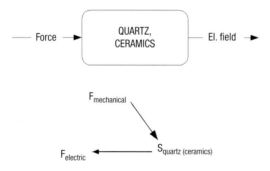

Fig. 3.10. Sufield models of the effect of piezoelectricity.

other applications. The effect is reversible; the piezoelectric crystals change shape when subjected to an externally applied electric filed.

    Magnetostriction (Fig. 3.11) is the property of ferromagnetic materials to change their geometric dimensions in a magnetic field. Conversely, when a ferromagnetic material

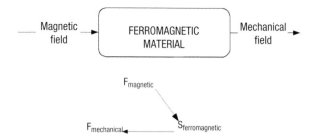

Fig. 3.11. Sufield models of the effect of magnetostriction.

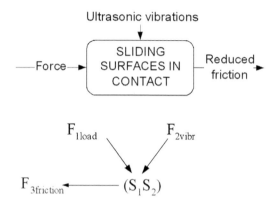

Fig. 3.12. Sufield models of the friction force reduction in an ultrasonic field.

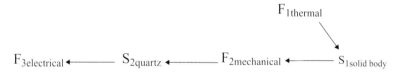

Fig. 3.13. Compound physical effect (thermal expansion + inverse piezoelectricity).

is under mechanical stress, it changes its magnetization. These effects are also used in various actuators (e.g., ultrasonic generators) and sensors.

The contact friction between two sliding parts decreases in the presence of ultrasonic vibration (Fig. 3.12). This effect is used to reduce loads in drawing, rolling, and other metal forming processes.

As one would expect, individual physical effects can form chains, such as the one shown in Fig. 3.13. This chain can be used, in one instance, to measure or detect heat-induced deformations of a metal part.

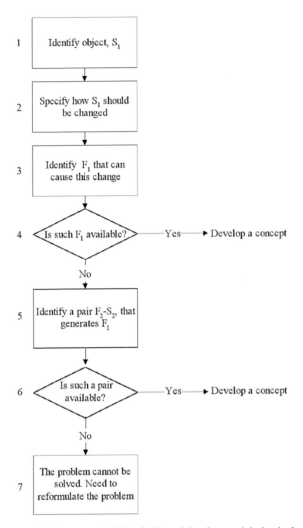

Fig. 3.14. Steps to build sufield models of potential physical configurations of new systems.

Over 6000 physical effects are known at the time of writing, and scores of new effects are being discovered all the time. The most comprehensive listing of many of these effects is available via software products from Invention Machine Corporation.

3.3   Using sufield models of physical effects

Sufield modeling of physical phenomena is very effective when used for the identification of potential physical configurations of new technological systems (Fig 3.14).

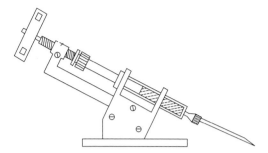

Fig. 3.15. Conventional microsyringe pump.

Problem 3.1

Microsyringe pumps are used for the infusion or withdrawal of small amounts of fluids into or from various biological objects (tissue samples, individual cells, etc.). A typical microsyringe pump (Fig. 3.15) includes a needle, container and piston. The piston is displaced manually (e.g., with the help of a micrometer knob) or by a computer-controlled electric drive. Modern microsyringe pumps can handle fluids with an accuracy of ± 0.01 μl. Let us use the steps in Fig. 3.14 to identify some potential physical configurations of a novel microsyringe pump.

(1) *Identify an object, S_1*:

This is the liquid to be displaced, $S_{1\text{liquid}}$.

(2) *Specify how S_1 should be changed*:

$S_{1\text{liquid}}$ should be moved out of a container.

(3) *Identify F_1 that can cause this change*:

This should be a mechanical field (force), $F_{1\text{mechanical}}$.

(4) *Is such a field available?*

Let's assume that the answer is "no."

(5) *Identify a pair $F_2 \rightarrow S_2$ that can generate $F_{1mechanical}$*:

At least three such pairs lead to alternative concepts.

- $F_{\text{thermal}} \rightarrow S_{\text{solid body or fluid}}$ (the effect of thermal expansion).
- $F_{\text{electrical}} \rightarrow S_{\text{piezoceramics}}$ (the effect of piezoelectricity).
- $F_{\text{magnetic}} \rightarrow S_{\text{ferromagnetic}}$ (the effect of magnetostriction).

(6) *Develop concepts*:

- $F_{\text{thermal}} \rightarrow S_{\text{solid body or fluid}}$: the liquid is displaced due to the expansion of a heated element.
- $F_{\text{electrical}} \rightarrow S_{\text{piezoceramics}}$: the liquid is displaced due to the expansion of a piezoceramics in an electric field.
- $F_{\text{magnetic}} \rightarrow S_{\text{ferromagnetic}}$: displacement due to the expansion of a ferromagnetic material in a magnetic field.

(Selection of the best concept is outside of the scope of this exercise.)

If, hypothetically, there were no pair $F_2 \rightarrow S_2$ that was capable of generating the required $F_{1mechanical}$, then one would need to solve a different problem: for example, *How to achieve the required result without injecting the liquid.*

3.4   Typical sufield transformations: the Standards

Numerous changes in technological systems can be described by several *typical sufield models* and their combinations. Most typical sufield models, described in this chapter, are presented in Table 3.3 (for simplicity's sake, the number and directions of actions in some of these models are not specified).

If a system, which needs a certain improvement, is modeled as a typical sufield, then a corresponding *typical sufield transformation* can be used to improve the system. Typical sufield transformations are called *Standard Approaches to Solving Problems* or, simply, *the Standards.*

The use of the Standards (generally, sufield analysis of any situation, Fig. 3.16) begins with abstracting an *initial sufield model* (*ISM*) from a freely worded description of a problem or situation (depending on one's understanding of the problem, more than one ISM may be compiled). At this point, the following have to be identified.

- Function (physical action – *see* Chapter 2 – to be performed).
- Object, S_1.
- Tool, S_2 (if it is present).
- Field(s), F (if any).
- Interactions between S_1, S_2, and F.

After the ISM's have been built, they must be associated with the available typical sufield models. These sufield models, in turn, must be related to one or more of the Standards that "prescribe" transition to more advanced, *desirable sufield models* (*DSMs*). This should be followed by determining the physical structures of the DSM's, and by the development of respective conceptual designs.

Appendix 3 contains all the known Standards (the *System of Standards*) and an algorithm for using them.

In the following sections, the most frequently used Standards are described.

3.4.1   Synthesis of complete sufields

The primary approach to the synthesis of sufields directly follows the definition of a minimal technological system.

Standard 1.1.1
To enhance the effectiveness and controllability of an incomplete sufield, it must be completed by introducing the missing elements (Fig. 3.17).

Table 3.3. *Typical sufield models*

The Standards are formulated according to a pattern: "if *(a certain condition or set of conditions)* is present, then *(a specific approach is recommended)*."

TYPICAL SUFIELD MODEL	TYPE OF SUFIELD
S_1 $\overset{F}{\diagup}$ S_1	Incomplete sufields.
$S_1 \overset{F_1}{-\!\!-\!\!-} S_2$ (with F_1 triangle over $S_1 - S_2$)	Complete sufield.
$S_1 - S_2$ with F_1 above and F_2 below (double diamond)	Double sufield.
$S_1 - S_2 - S_3$ with F_1 and F_2 triangles	Chain sufield.
$S \overset{F_1}{\diagup} \,\, \overset{\searrow}{F_2}$; $S_1 - S_2$ with F_1 and F_2	Detection/measurement sufields.

The completion of sufields is usually employed when introducing a new useful action or when improving an inadequate one (*see* Chapter 2, Fig. 2.5).

Problem 3.2

Hot rolling of metals requires the lubrication of the deformation zone (Fig. 3.18). Usually, the liquid lubricant is applied by special brushes and sprayers. Neither technique provides a uniform distribution of the lubricant in the deformation zone; the oil splashes around, and much of it is lost. A cost-effective method for the uniform supply is needed.

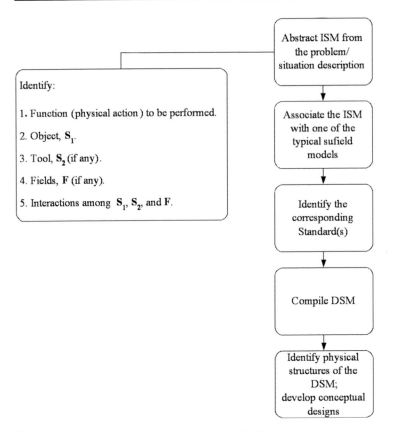

Fig. 3.16. Building sufield models using the Standards.

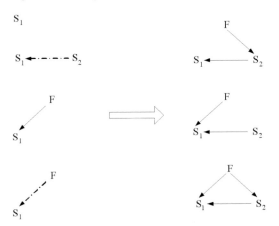

Fig. 3.17. Incomplete sufields should be completed.

Analysis

The primary function in this problem is the delivery of the required amount of lubricant to the deformation zone. The lubricant can be considered an object, $S_{1lubricant}$. The problem emerged due to the absence of a proper tool that would accurately carry

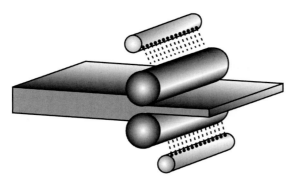

Fig. 3.18. Conventional lubrication of the deformation zone.

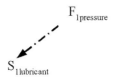

Fig. 3.19. Initial sufield model of the conventional lubricant supply process.

Fig. 3.20. DSM for a new lubricant supply process.

$S_{1lubricant}$. Also, in the original system, a mechanical field (hydraulic pressure), produced by the lubricant sprayers and acting on $S_{1lubricant}$ is inadequate.

Given these considerations, one can compose the following ISM (Fig. 3.19).

According to Standard 1.1. this ISM should be completed by introducing a pair $F_2 \rightarrow S_2$ (Fig. 3.20) that would exert a mechanical force on $S_{1lubricant}$.

Now we ought to identify S_2 and F. An ideal S_2 would carry the lubricant, and not interfere with the rolling process. This implies that S_2 should disappear in the deformation zone upon delivery of the lubricant. As for F_2, it would be ideal to use resource fields available in the system. These are the mechanical field of rotating rollers and the thermal field of the deformation zone.

Solution

This discussion helps us to arrive at a practical solution: A paper impregnated with the liquid lubricant is fed between the rollers and the metal strip. The paper is burned due to the high temperature in the deformation zone (Fig. 3.21).

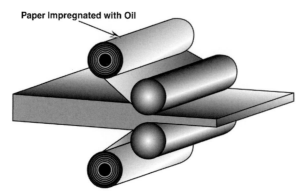

Paper Impregnated with Oil

Fig. 3.21. Paper delivers the lubricant to the deformation zone.

As one can see, a sufield analysis does not by itself generate specific answers to problems, but rather helps to conceive of an array of conceptual solutions. Subsequent engineering and cost analyses of these alternatives allow one to select the best solution.

3.4.2 Elimination of harmful actions in sufields

The following group of Standards resolves situations (system conflicts) when two substances in a complete sufield experience, both useful and harmful interactions with each other. The task is to retain the useful interaction while eliminating the harmful one. Frequently, this can be achieved by introducing a new, third substance, S_3, that partitions the original substances.[1]

Standard 1.2.1

If both useful and harmful actions develop between two substances, and the use of outside substances is allowed, then a third substance should be introduced between the two (Fig. 3.22).

The added substance should be as inexpensive as possible, preferably made from readily available resources.

Problem 3.3

To conduct a scientific experiment, a set of a few thousand precisely dimensioned 0.01-mm-thick round aluminum disks (made from aluminum foil) was required (Fig. 3.23). However, the turning on a lathe of a soft stack of thin aluminum disks would result in their bending, and the tearing of their edges.

---

[1] Basically, this and other Standards on elimination of harmful actions are sufield interpretations of the third approach for resolving system conflicts: introduction of a new tool (*see* Chapter 2). These interpretations help one to identify a new auxiliary tool being introduced into the system either as a substance or a field.

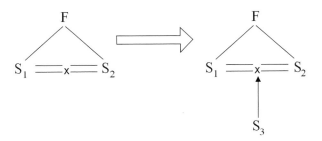

Fig. 3.22. Introducing an outside substance as a partition.

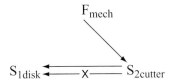

Fig. 3.23. How to machine the slender disks?

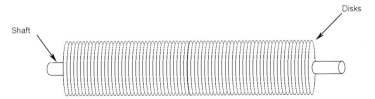

Fig. 3.24. ISM for the disk machining problem.

Analysis

Initial sufield model (Fig. 3.24) comprises S_1 (a disk), S_2 (the cutter), and F_{mech}. In this ISM, cutting the disk is the useful action, while damaging it is the harmful one.

According to Standard 1.2.1, the situation may be improved if S_1 is combined with another substance, S_3, that would make it stronger for the period of machining (Fig. 3.25).

Solution

The problem is solved by stacking the blank disks in a water-filled container and freezing the water. The hard ice block containing the blanks is easily machined.

Often, however, the introduction of a "foreign" substance may excessively complicate the system. Thus, a system conflict develops: in order to break a harmful action, a third substance, S_3, has to be introduced, but this would overcomplicate the system.

This system conflict can be resolved by making the third substance out of the original substances, according to the following Standard:

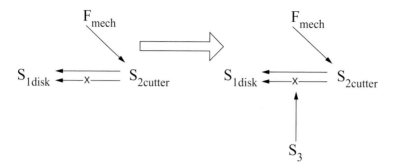

Fig. 3.25. Introducing S_3 to reinforce the disk(s).

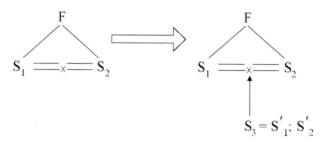

Fig. 3.26. Added substance is a modification of the original substances.

Standard 1.2.2

If both useful and harmful actions develop between two substances, and the use of outside substances is prohibited or undesirable, then a third substance, which is a modification of at least one of the original substances, should be introduced between the two (Fig. 3.26).

To produce modified substances, several approaches can be used. It may be possible to change the aggregate state of the available resource substances. Ice and steam, for instance, can be derived from water and therefore can be considered its modifications. Another way is to decompose the given substances into their constitutive elements (e.g., splitting water molecules into atoms of oxygen and hydrogen). A modified substance can also be obtained by changing the physical state of the original substance, e.g., by magnetization, ionization, making a stationary object move or stopping a moving object, changing the temperature, combining two original substances, etc.

Introducing S_3 (as well as any other substances or fields) into sufields may require constructing *auxiliary sufields* (such as shown in Fig. 3.27), whose purpose is to control S_3 (e.g., to create it, move it, hold it, etc.).

Problem 2.10 (the wear of pipes in shot blasting machines) is a good illustration of this approach (Fig. 3.28).

When performing sufield analysis, the main sufields (which include S_1 and S_2) should be constructed first. If auxiliary sufields are needed, they can always be built later in the

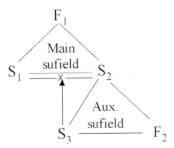

Fig. 3.27. Auxiliary sufield controls the added substance.

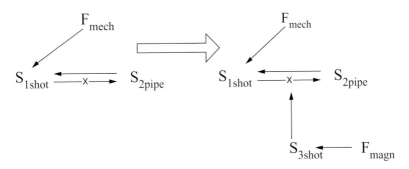

Fig. 3.28. Sufield solution of the shot blasting problem.

process, after S_3 has been identified. For simplicity, the auxiliary sufields are usually not shown in sufield models.

Standard 1.2.2 is one of the most frequently used. Consider another example.

Problem 3.4

Hydrofoil ships are often used for both military and commercial applications (Fig. 3.29). The hydrofoil lifts the ship's hull out of water, thus reducing the drag. The hydrofoils are subjected to intense cavitation erosion. Cavitation is the formation of gas bubbles (cavities) in the water at the moving solid body. The bubbles collapse inwardly, thus creating instant and very high-pressure changes (shock waves) that damage the hydrofoil surface. If the cavitation is intense, the hydrofoil surface is rapidly covered with numerous pits and perforations. Making the hydrofoils from refractory metals or using various protective layers produces a marginal effect.

Analysis

In this problem, a hydrofoil is S_1, water is S_2, and the field is $\mathbf{F_{mech}}$, which is a mechanical force propelling the ship. The conditions of the problem make it clear that partitioning the water and the hydrofoil with a different substance will not achieve the desired result. The problem is analogous to the previous one, and so is its solution: S_3 should either be a modified hydrofoil or modified water (Fig. 3.30). Since modifications

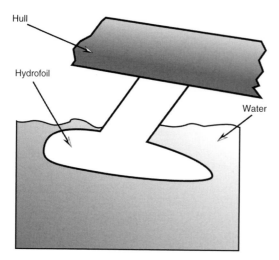

Fig. 3.29. Hydrofoil in water.

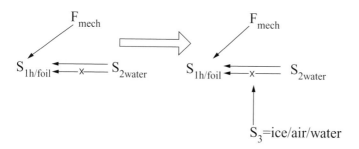

Fig. 3.30. Sufield solution of the hydrofoil erosion problem.

of hydrofoil were already attempted, one should try modifications of water. These may be steam, ice or water with altered parameters, i.e. different speed and/or pressure.

Solution
Two solutions are practiced: (a) The forming of an air curtain around the hydrofoil by pumping air through perforations in its leading edge and (b) forming and maintaining a thin layer of ice on the hydrofoil surface (e.g., by a small refrigeration machine on the ship); this layer is continually destroyed by cavitation, but is immediately recreated.

Sometimes, the conditions of a problem ban any partitioning of S_1, and S_2. Then, the following Standard applies (Fig. 3.31).

Standard 1.2.4
If both useful and harmful actions develop between two substances, and a direct contact between the substances should be maintained, then the useful action is provided by the existing field F_1 while a second field, F_2, neutralizes the harmful action.

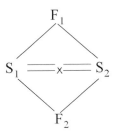

Fig. 3.31. Second field neutralizes/eliminates the harmful interaction.

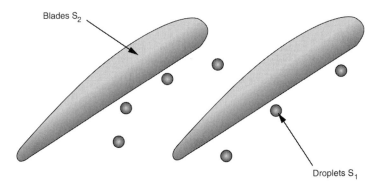

Fig. 3.32. Droplets damage the blades.

Problem 3.5
Steam turbines operate in the environment of a mixture of steam and water droplets. Relatively large drops (typically 50 to 800 μm in diameter) collide with the quickly rotating blades and cause their erosion (Fig. 3.32).

Sufield analysis
From a sufield analysis standpoint, this situation is somewhat similar to the previous problem of hydrofoil erosion.(Fig. 3.33). However, a similar solution – forming a protective layer of ice on the blade surface – is hardly realistic. This stipulates the application of Standard 1.2.4 (Fig. 3.34). The task for a new field, F_2 is clear: to prevent the contact between the blade surface and the droplets. Consulting basic physics leads to the idea of using an electrostatic field.

Solution
The same electric potential is applied to the droplets (using a corona discharge – the technique employed in electrostatic filters) and to the turbine blades (Fig. 3.35). Thus, the droplets and the blades repel each other, effectively reducing the relative velocity with which the droplets impact the blades.

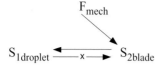

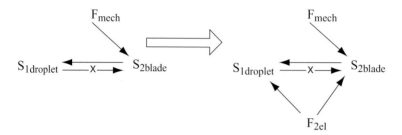

Fig. 3.33. IMS for the blade erosion problem.

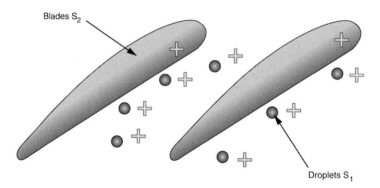

Fig. 3.34. DSM includes a second field.

Blades S_2

Droplets S_1

Fig. 3.35. Charged droplets and blades repel each other.

3.4.3 Detection and measurement sufields

There are two types of problems encountered in engineering practice: *modification* problems and *measurement/detection* problems. In problems of the first type, the functioning of a given system should be improved. This is commonly achieved by modifying either the system, or its environment or both. The preceding problems have all been modification problems.

In measurement/detection problems, a system whose parameters or physical state have to be gauged is given, and it cannot be changed. The first approach to solving problems of measurement and detection is . . . to withdraw from solving them. In most systems, measurement and detection are important, but nonetheless complementary

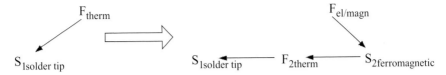

Fig. 3.36. Sufield solution of temperature stabilization problem.

functions. The necessity to measure/detect some parameters of a system is indicative of certain flaws in the system's performance (i.e., the system cannot itself maintain these parameters within the specified range and, therefore, needs controllers of some sort). Ideally, a system should function so that the need for measurement/detection of its parameters does not exist. This approach is manifested in the following Standard:

Standard 4.1.1
If there is a problem requiring measurement or detection in a system, then the system should be changed so that the need to measure or detect it is eliminated.

Problem 3.6
Soldering irons are used to rework and repair expensive printed circuit boards. The iron's tip temperature should be minimally acceptable to avoid damage to the components and boards. Since various jobs require different temperatures, the irons must be constantly recalibrated, which reduces the overall productivity – and annoys the technicians. Also, the technicians often increase throughput by setting the irons to a higher tip temperature than is minimally required. This significantly increases the risk of overheating the components and boards. Using automatic temperature control systems would add complexity and cost.

Analysis
It would be ideal if a complex system for precision temperature control would "convolve" into a substance capable of maintaining temperature within the specified range. To accomplish this, it is natural to use the transition through the Curie point, which is the temperature at which a ferromagnet loses its magnetic properties (Fig. 3.36).

Solution
In the soldering iron made by Metcal, Inc., the tip is a part of an interchangeable cartridge. It is a copper bar coated with a thin layer of an iron–nickel alloy (Fig. 3.37). A high-frequency alternating magnetic field, produced by the surrounding coil, induces eddy currents in the iron–nickel layer – thus heating it up. When the temperature of the layer reaches the predetermined Curie point, the layer becomes impermeable to the magnetic field, diverting the current to the internal copper core. As a result, the

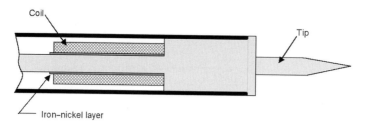

Coil

Tip

Iron–nickel layer

Fig. 3.37. Interchangeable cartridge with a self-regulating soldering tip.

intensity of induction heating drops. When the layer cools, it becomes permeable again. The current returns, and the heating cycle continues. The stability of this soldering iron is $\pm 1.1°C$. Different tip temperatures are achieved by using iron–nickel alloys with different Curie points.

In situations when Standard 3.1.1 is not applicable (e.g., a system that does not readily submit to change), help comes from the concept of the complete sufield.

In problems of measurement/detection, an object of interest can be considered a substance S_1. From a sufield analysis standpoint, to be measured or detected S_1 should generate some output field that carries information about the substance's condition (Fig. 3.38a).

If S_1 cannot generate such a field by itself, another field is applied to S_1 so that the substance transforms this input field into the needed output field; parameters of the latter depend on the physical state of S_1 (Fig. 3.38b).

If S_1 does not respond to a certain field that may be of practical interest, or conditions of the problem restrict the use of this field, then a second substance, S_2 is added. This substance can either change the parameters of an input field (Fig. 3.38c), or it can transform an input field into the different output field (Fig. 3.38d), or it can generate the required output field itself (Fig. 3.38e). The sufield groups shown in Figs. 3.38c and 3.38d are the most typical.

Thus, measurement and detection of technological systems involves the building of sufields, with those output fields that can easily carry information. To synthesize a measurement/detection system, it is recommended to use the approach similar to synthesizing modification systems, i.e., completion of the sufield. A distinguishing feature specific to the synthesis of measurement/detection sufields is that the sufield must assure that an output field is obtained.

Standard 4.2.1

If it is difficult to measure or detect elements of an incomplete sufield, then this sufield has to be completed and have a field as its output.

Of all the fields available to modern technology, electromagnetic fields (i.e., magnetic, electric, optical) are the most convenient to use, since they can be easily measured, detected or transformed into other fields and transmitted in a three-dimensional space.

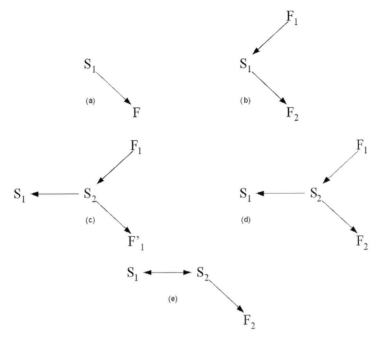

Fig. 3.38. Typical measurement/detection sufields.

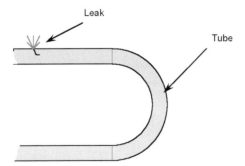

Fig. 3.39. How to detect a leak in the refrigerator tubing.

Problem 3.7

Refrigerators are tested against potential leaks of a cooling agent in refrigerator tubing. A simple, cost-effective and reliable method for visual detection of leaks is needed (Fig. 3.39).

Sufield analysis

A leaked droplet can be considered as S_1. The ISM is an incomplete sufield. A new S_2 should be introduced. It would either interact with an optical field, $\mathbf{F_{optical}}$, in such a way that the droplet becomes visible, or it would transform some other field into an optical field, $\mathbf{F_{optical}}$ (Fig. 3.40).

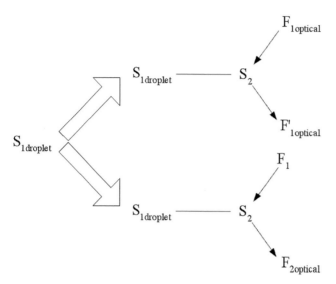

Fig. 3.40. Sufield models for detecting the droplet.

Solution

Both of these approaches are used. The first approach is realized with the help of a layer of a paint that is applied to the external surface of the refrigerator tubing. If leakage occurs the paint swells, making the visual inspection easier.

The other approach utilizes the ability of luminophores to convert UV radiation into visible light. Adding a minute amount of luminophore into the cooling agent will make UV-irradiated leakage spots glow.

3.5   How to introduce new substances into sufields

The basic concept of the sufield directs one to complete fragments of sufields by introducing missing substances and/or fields into the existing systems. Frequently, this may result in increasing the complexity of the systems, and even cause undesirable technical effects. Thus, while constructing a sufield, one may face the conflict: construction of an effective sufield (system) requires the introduction of some substances and/or fields, but this may excessively complicate the system. Such conflicts can be resolved by using various indirect ways to "introduce substances/fields without their introduction."

The introduction of substances can be circumvented by the following approaches.

Standard 5.1.1

If a substance needs to be added to the system, but it is unacceptable due to the problem specifications or the performance conditions, then some indirect ways should be used.

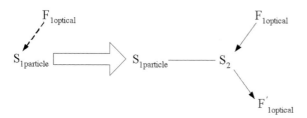

Fig. 3.41. Sufield analysis of the particle detection problem.

- *Use a substance that is already a part of the system's environment.*
  Example: Problem 2.4 of the self-unloading barge in Chapter 2.
- *Use voids instead of substances.*
"Void" does not necessarily mean "vacuum." The "void" in a solid substance can be a pore or a capillary, but also a liquid or a gas. In a liquid, a void can simply be a gas bubble. A void is an exceptionally useful resource. It is usually available, unlimited, extremely cheap, and can be easily "mixed" with other resources forming, for instance, hollow and porous structures, foams, bubbles, etc.

Problem 3.8
Manufacturers of optically clear liquids should check them for impurities. If a sample of a liquid contains more than a certain number of impurities, the whole volume of that liquid should be reprocessed. Impurities are small, non-magnetic particles of various natures that may enter the liquid at different stages of the production process. The preferred detection method is optical: a laser beam scans a sample of the liquid, it bounces off the particles and the number of reflections is counted.

There are, however, particles so small (about or less than 580 nm in diameter) that they do not reflect light well (the light wavelength is comparable with their size, so light "flows" around them).

It is necessary to devise an *optical* method for detecting (counting) small particles.[2]

Sufield analysis
The ISM of the problem includes a particle, $S_{1particle}$, which cannot interact with an optical field, $F_{1optical}$. A second substance, S_2, attached to S_1, should be added to transform $F_{1optical}$ into a reflected light, $F'_{1optical}$ (Fig. 3.41).

S_2 should satisfy three major requirements: (1) it should be big enough to reliably reflect the incident light; (2) it should be "sticky," to attach itself to the particle; and

---

[2] One is tempted to solve different problems, for example, *how to prevent the contamination of the liquid;* or *how to remove the particles from the liquid without detecting them;* or *how to detect the particles using non-optical means.* Problem 3.8, however, purports to illustrate a certain TRIZ approach; a full-blown analysis of all potential solution paths is beyond its scope.

(3) it should not contaminate the liquid. It is relatively easy to meet requirements 1 and 2, but the last requirement is a real challenge. Ideally, the liquid itself should be modified to serve as S_2.

Solution

Gas bubbles are used to detect the particles. The liquid in a sample volume is heated: if the particles are present in the liquid, the liquid around them begins to boil, producing small bubbles (Talanquer, 2002). The bubbles grow rapidly and are easily detected.

• *Use a modification of the object or its part (ingredient) as a tool.*

Problem 3.9

Pure crystals of beryllium oxide are produced by growing them from a beryllium oxide melt. For this process, the powder of beryllium oxide has to be heated in a contamination-free environment up to a temperature exceeding 2000°C. Heating of the oxide in a crucible, made from refractory metals, does not guarantee that some atoms of the crucible material would not get into the melt. The situation contains an apparent physical contradiction: a melting pot must be present to contain the melt, and the pot must be absent to prevent contamination of the melt.

Analysis

To resolve this contradiction, it was suggested to melt the beryllium oxide within . . . the beryllium oxide itself. If one could heat the center of the beryllium oxide powder with a high frequency current (induction heating), then the melted portion of the powder would not be exposed to anything other than the oxide. The oxide is dielectric, however, and becomes conductive only in the molten state. On the other hand, the introduction of a conductor (it may be particles of some metal) into the powder may cause the contamination of the oxide.

This is an exemplary situation: to solve a problem, one has to overcome not one, but many barriers, a so-called "contradiction chain." The second physical contradiction: a conductor must be present in the powder to allow for induction heating, and a conductor must be absent so as not to contaminate the oxide. Common sense prompts the idea of using a modification of the beryllium oxide as the conductor.

Solution

Metallic beryllium is used as the conductor. This assures the absorption of the induction field and the heating of the oxide mass. At the high temperature, however, the beryllium burns and oxidizes, and therefore does not contaminate the melt.

• *Obtain a substance from a chemical compound.*

Problem 3.10

There are lubricants (Malinovsky, 1988), which are colloidal suspensions of fine metal particles in oil. In operation, the particles precipitate from the solution and are deposited on the rubbing surfaces, thus forming a protective layer. The lubricants are usually made by first crushing the metal powder in a grind mill, then mechanically mixing it with the oil.

Although very effective, such lubricants cannot be used in situations when the metal particles in the solution are bigger than the clearance between the two rubbing surfaces. More intensive and lengthy grinding of the powder might produce somewhat finer particles, but this would increase the manufacturing costs.

Sufield analysis
The perfect way to use the metal particles might be to switch from a colloidal suspension to a true solution, in which the metal would be present in the form of molecules or atoms. Metals, however, do not normally dissolve in oils. The physical contradiction is clear: the metal should be present in the lubricant in order to plate the surfaces, and the metal should not be in the lubricant because it cannot dissolve in the oil. This is a typical problem in introducing a substance, where the problem specifications or performance conditions of the system make the introduction look impossible.

Solution
A chemical compound containing the required metal is used. Such a compound has to satisfy two demands: it must dissolve in oil, thus creating a true solution, and it must liberate the metal under the effect of the friction forces. Since oil is an organic substance, using organo-metallic compounds can fulfill the first demand.

The second demand is satisfied automatically, because these compounds decompose in a thermal field generated by the rubbing surfaces.

As one can see, the last two approaches are opposites of each other. While the former suggests extraction of the required substance from a compound, the latter recommends the use of a simple substance that should, upon completing its function, become a part of a compound.

• *Use a field instead of a substance.*

Problem 3.11
It is necessary to monitor both the stretching of the track cable and the speed of the gondola in cable car installations (Fig. 3.42). The accuracy of any proposed method should not be affected by various environmental factors, such as temperature, moisture, soiling of the cable, etc.

Sufield analysis
The original measurement sufield is shown in Fig. 3.43. Here, a "marker" is any potential substance (e.g., paint). The system conflict is obvious: an external "marker" provides the measurement of the cable's conditions, but it is damaged by the cable's environment.

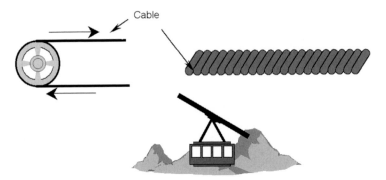

Fig. 3.42. Cable car.

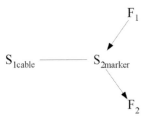

Fig. 3.43. IMS for the cable measurement problem.

Solution
This system conflict is resolved in the spirit of ideality tactic 2 (Chapter 2): the "marker" is moved inside the cable.

Magnetic marks are used instead of paint (Fig. 3.44). The pulse length conveys information about the degree of cable stretching, while the frequency of pulses is proportional to the cable's speed. The magnetized sectors of the cable are much less affected by the environmental factors, thus increasing both the accuracy and the reliability of the monitoring process.

3.6   How to introduce new fields into sufields

Similarly to the introduction of substances, the introduction of fields follows the same general principle of delegating the function to one of the existing substance or energy resources.

Standard 5.2.1
Use fields readily available in the system.

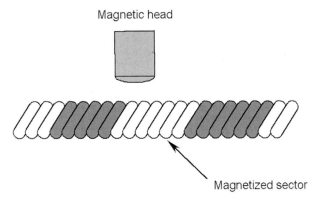

Magnetic head

Magnetized sector

Fig. 3.44. Magnetic marks allow for measuring the cable.

$S_{1bubble}$ —————— $S_{2liquid\ oxygen}$

Fig. 3.45. ISM for the bubble removal problem.

Fig. 3.46. Centrifugal force removes the bubbles.

Problem 3.12

When a rocket tank is being filled with liquid oxygen, it is necessary to keep bubbles of oxygen gas from entering the tank.

Sufield analysis

The ISM consists of two substances, $S_{1bubble}$ and $S_{2liquid\ oxygen}$ (Fig. 3.45). There is no field that could control their separation. (Note that the buoyancy force (i.e., the gravitational field) is not included in the ISM because it does not submit to control.)

An easily controllable field should complete the sufield. This field must act differently on the substances, thus causing their separation. The dissimilar interactions between the field and the substances can be supported by a certain contrast in the physical properties of the substances, i.e., by the difference in their masses. Then, it would be natural to use a centrifugal force as a field of separation (Fig. 3.46).

The introduction of a centrifugal force should not result in the overcomplicating of the system. This suggests putting some resource fields into service, first – a mechanical field of the stream of the liquid oxygen.

Solution

To solve the problem, it is enough to transform motion patterns of both the liquid oxygen and the gas by swirling the stream. Spinning can be accomplished by placing a plate inside the pipe that is slightly inclined toward the axis of the stream. The centrifugal force presses the coolant to the walls and pinches the gas towards the axis of the pipe. The gas is then removed from the stream by various means, i.e., with the help of a catheter.

Standard 5.2.2

If internal fields are not available, use fields present in the system's environment.

Most present-day particulate filters used in automotive diesel engines require servicing (cleaning) every 75 000 miles. A new particulate filter, developed by PSA Peugeot Citroen (http://www.psa-peugeot-citroen.com/en/morning.php), self-cleans itself every 200–300 miles.

The filter consists of a capillary structure, made of silicon carbide that is covered with a catalytic coating. Self-cleaning involves the periodic burning-off of the particulates that have accumulated in the filter. Sensors check the temperature and the pressure within the particulate filter to notify a fuel injection computer about the amount of soot collected in the filter's pores. The fuel injection computer initiates the post-injection of fuel after the main injection has taken place. The exhaust gas temperature rises over the particulate combustion threshold ($550°C$), the overheated gas sweeps through the filter and burns the soot.

Standard 5.2.3

If neither internal nor external fields are readily available, then use fields whose carriers or sources are the substances already present in the system or in the system's environment.

Problem 3.13

A thermal sensor is used to detect possible overheating of the sliding bearing that supports a rotating shaft of an expensive mechanism (Fig. 3.47). In case of overheating, the sensor sends an electric signal to a safety switch that immediately stops the rotation of the shaft. The heat flux from the bearing may take unacceptably long to reach the sensor, thus delaying the switch's response and endangering the mechanism.

Sufield analysis

This is a typical problem in constructing a measurement sufield. The ISM (the left part of the diagram in Fig. 3.48) consists of $S_{1bearing}$ and $F_{thermal}$. This is an incomplete sufield, so the missing elements have to be added. The DSM (the right part of the diagram) shows a complete sufield in which a thermal interaction between $S_{1bearing}$

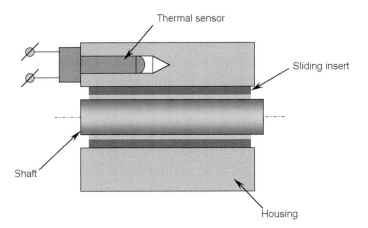

Fig. 3.47. Measuring the bearing temperature.

Fig. 3.48. Sufield model for measuring bearing's temperature.

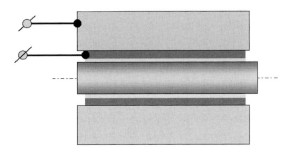

Fig. 3.49. Thermal couple composed of the resource substances.

and S_2 produces a preferable electric field, F_{el}. Thus, the problem is reduced to finding such a substance S_2 that, jointly with $S_{1bearing}$, would generate an electric field.

Standard 5.2.3 suggests that this substance has to be found among the resources of the existing systems.

It was suggested to use the housing as S_2. It is well known that a difference of electric potentials is generated between two conducting materials whose temperatures are different when in contact (the *Seebeck effect*[3]). This phenomenon was employed

---

[3] It is the reverse of the *Peltier effect* (a current flow causes a temperature gradient across the joint of different metals, Nolas *et al.*, 2001).

to solve the problem. The thermo-electric voltage generated by the housing and the sliding insert control the safety switch (Fig. 3.49).

3.7   Summary

The sufield is a model of a technological system that reflects one (preferably, the most important) physical property of the system given in the problem. As a model, the sufield is highly universal; it allows for describing a broad range of technological structures and their changes in terms of substance–field interactions. The main purpose of improving the sufield structure of a system is to increase the system's controllability. A properly constructed sufield model reflects the essence of a problem, and does not include secondary, for the given problem, effects. The sufield modeling approach provides one with the means to record *what is given* in the problem and *what should be obtained*.

3.8   Exercises

3.1 Compile sufield models of the following functions:
- sitting
- brushing teeth
- combing hair
- skiing
- boiling water in a kettle
- lighting the bulb in Fig. 3.50.

3.2 The flaring of metal pipes is often performed by special hydrostatic presses (Fig. 3.51). However, this technology is not always available for emergency repairs in field conditions. A simple and cost-effective method for tube flaring is needed.

3.3 Dirt often sticks to the inner surface of an excavator bucket during unloading, which is undesirable (Fig. 3.52). Suggest a simple and cost-effective method for eliminating this disadvantage.

3.4 Various loose materials are often transported in partially filled containers (Fig. 3.53). During transportation, the material is subjected to sloshing vibrations that may cause particles to be frequently displaced, and to rub against both each other and the walls of the container. This may result in a dusting of the particles, and a build-up of static electricity that may cause an explosion. Suggest conceptual solutions that would prevent these effects.

3.5 In the paper industry, logs provide the raw material for the manufacture of paper products. Logs delivered to a mill often include pieces of plastics (the consequence of environmental pollution). When the logs are converted into wood chips, the plastic is fragmented as well, which results in the contamination of the paper

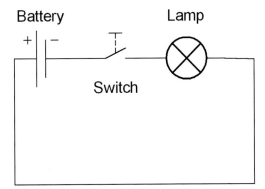

Fig. 3.50. Circuit for lighting a bulb.

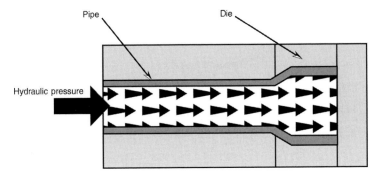

Fig. 3.51. Hydrostatic method of pipe flaring.

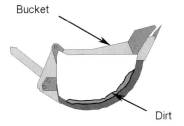

Fig. 3.52. Excavator bucket.

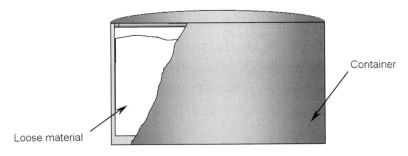

Fig. 3.53. Transporting loose materials in containers.

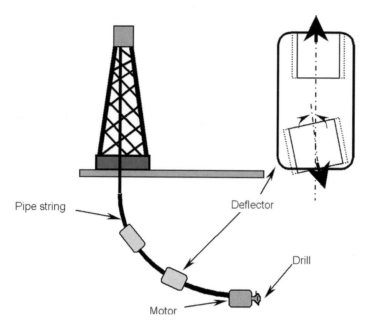

Fig. 3.54. Drilling curved wells.

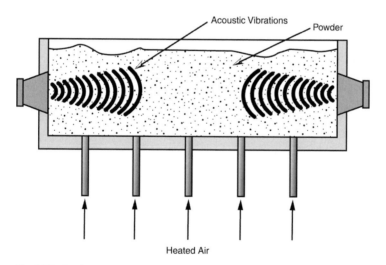

Fig. 3.55. Drying powders in a fluidized bed.

products. The prevalent practice is to visually inspect the logs and remove the visible pieces of plastic. It is needless to argue about the low efficiency of this method. Develop a better method for plastic removal.

3.6 The rotary drilling rig uses a series of rotating pipes, called the drill string, to tap into oil and gas reservoirs (Fig. 3.54). The drill string is supported by a derrick, and driven by the rotary table on its floor. A mud-like fluid, circulated by a pump, removes cuttings as the teeth of the drill bit dig into the rock around the oil/gas reservoir.

Special deflectors are used to drill curved wells. As the radius of curvature of the well changes, new deflectors have to be switched onto the drill string. To accomplish this, the whole drill string has to be lifted, and then lowered back into the well. This dramatically reduces the overall effectiveness of the drilling process.

Suggest conceptual designs for the deflectors, free of the above shortcoming.

3.7 When drying powders in a fluidized bed (Fig. 3.55), a layer of powder is heated by a stream of hot air while being agitated by acoustic vibrations. The agitation intensifies the drying process. The best result can be achieved if the frequency of the acoustic vibrations is tuned to the natural frequency of the particles. Since the particles are small, their natural frequency is in the ultrasonic range. However, ultrasound propagates poorly in powders. Formulate a mini-problem, and solve it using sufield analysis.

4 Algorithm for inventive problem solving (ARIZ)

The tools of TRIZ, as described in Chapters 2 and 3, are very helpful when searching for new conceptual solutions. Their direct application to complex situations, however, can at times be rather perplexing: it may not be clear what system conflict should be resolved, or which component of the system should be modified, or what sufield model describes the problem most adequately. To assist the problem solver in making these and other important decisions, Altshuller developed the *Algorithm for Inventive Problem Solving* or *ARIZ (Algoritm Reshenia Izobretatelskih Zadach* – in its Russian abbreviation).

4.1 Goals of ARIZ

ARIZ[1] pursues three major objectives: *problem formulation, breaking psychological inertia*, and *combining powers of various tools of TRIZ.*

4.1.1 Problem formulation

A conventional approach to problem solving is to jump to potential solutions once a problem has been presented. Solving a difficult problem may require numerous leaps. From a TRIZ standpoint, if a problem cannot be easily solved, it usually means that it is poorly formulated. A properly formulated problem can be easily solved automatically, or it becomes clear that there is no necessary scientific knowledge or technology available to find a solution. In many cases, the proper problem formulation includes the selection of a system's component to be modified, and the formulation of a physical contradiction for this component.

For this reason, ARIZ contains a set of steps that lead the problem solver from the initial vaguely, or wrongly, defined problem to a lucid formulation of a system

---

[1] Genrikh Altshuller is the author of the concept of ARIZ and of its many versions: ARIZ-61, ARIZ-64, ARIZ-69, ARIZ-71, ARIZ-75, ARIZ-77, etc. (the numbers stand for the year of the development of a particular version). The present version is a modification of ARIZ-85C developed by Victor Fey.

conflict and a physical contradiction. Each step of ARIZ is intended to modify the initial understanding of the situation in such a way that arriving at a solution becomes much easier. Frequently, correct implementation of the initial steps of ARIZ may lead to a solution even without formulating a physical contradiction. Thus, ARIZ is mainly concerned with reformulating the initial problem; a solution should develop as an outcome of a correctly performed analysis.

4.1.2   Breaking psychological inertia

ARIZ in its present form is an algorithm designed for humans, not for computers, and therefore is equipped with a set of tools for breaking the psychological inertia of the problem solver.

The first tool is a recommendation to describe a problem without using the professional lingo, which reflects existing perceptions about the system and may erect high psychological barriers on the path to an effective solution.

The second tool is *intensification of system conflicts.* Conventional thinking does not explicitly recognize system conflicts or try to alleviate them. Contrary to that, ARIZ (as well as all of TRIZ) thrives on system conflicts. To offset one's natural tendency to soften system conflicts, ARIZ offers an opposite approach, that is, their aggravation. Driving system conflicts to their extremes helps the problem-solver to curb his or her own psychological inertia, and often enables a deeper understanding of the problem.

4.1.3   Combining powers of various tools of TRIZ

ARIZ combines all the basic notions and techniques of TRIZ such as the ideal system, system conflict, physical contradiction, as well as the sufield analysis, the Standards, and the laws of evolution. This makes ARIZ the most powerful tool for problem analysis and solution in modern TRIZ.

4.2   Structure of ARIZ

The four-part structure of ARIZ is shown in Fig. 4.1.
- Part 1 – *Formulation of system conflicts* – starts by specifying the primary function of the given system. Then, major components of the system and the actions they perform are identified. Next, objects, tools and *environmental elements* are specified. *An environmental element is a component that belongs to the system's environment or to an adjacent system, and that adversely affects the system's components* (e.g., oxygen in the air that causes undesirable oxidation). Tools are categorized as either main or auxiliary.

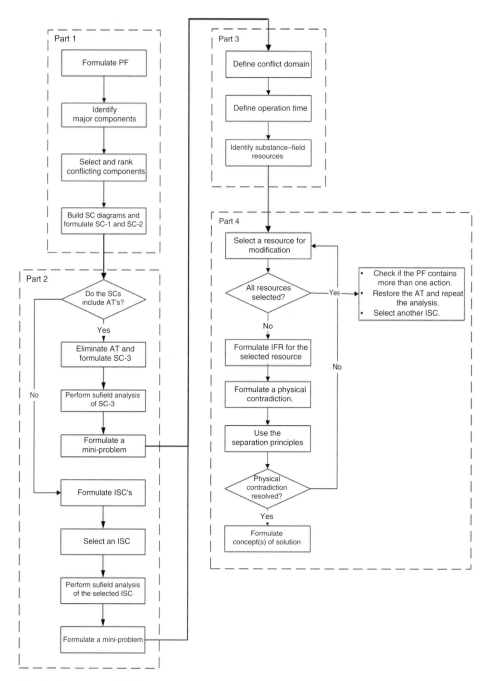

Fig. 4.1. Structure of ARIZ.

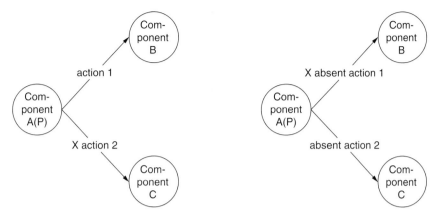

Fig. 4.2. System conflicts always go in pairs.

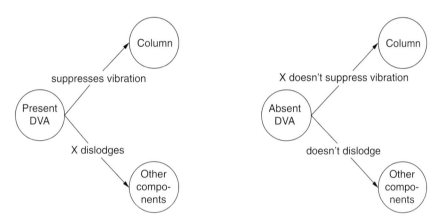

Fig. 4.3. Two physical states of the DVA define the two system conflicts.

ARIZ distinguishes two inverse system conflicts associated with the initial situation. The physical state of one of the system components determines its interaction with other components. If this component is in the opposite physical state, then the interaction will be inverted, too (Fig. 4.2). For example, the two physical states of a dynamic vibration absorber in Problem 2.3 define the two system conflicts diagrams in Fig. 4.3. The outcome of ARIZ analysis will depend on which system conflict will be resolved.

- Part 2 – *Analysis of the system conflicts and formulation of a mini-problem* – begins with the checkpoint: do the formulated system conflicts involve an auxiliary tool? As described in Chapter 2, if a system conflict involves an auxiliary tool, it may be resolved by eliminating this tool and delegating its useful action to another existing component, or to the system's environment. Thus, if the answer to this question is yes, the auxiliary tool is eliminated (along with all of its useful and harmful actions), and a new system conflict is formulated. Very often, the system conflicts involve an auxiliary tool in two physical states: present and absent. In this case the system conflict with the absent auxiliary tool is selected.

If the system conflicts do not contain an auxiliary tool, they are intensified: the opposing physical states or harmful actions of the main tool are driven to their imaginable extremes. In many cases, overcoming one of the intensified system conflicts will lead to solving a maxi-problem, while overcoming the other will be within the framework of the mini-problem approach. A set of rules helps the problem-solver select the most promising system conflict.

One of these rules deals with situations when the intensification of both system conflicts makes performance of the primary function impossible or causes significant harm to the overall system. In such circumstances, the main tool must be eliminated, and a new system conflict formulated. This rule is based on the view that intensified system conflicts help reveal the evolutionary potential of the system, or of its underlying physical principle. The critical deterioration of the system, due to any significant change of some main component's important parameter, is a reliable indication that this component has reached the bounds of its evolutionary development and that any further major improvement is most likely unattainable. A new component to perform the required physical action needs be introduced or a new physical principle of functioning (i.e., a new physical action) should be considered.

At the next step, sufield analysis helps to determine whether the selected system conflict can be resolved by one of the Standards. Even if the solution becomes clear at this step, it makes sense to continue the analysis since the following steps may help refine and reinforce the solution.

The final procedure in Part 2 is the formulation of a mini-problem. Summing up the previous steps, three outcomes of the system conflict analysis (and correspondingly three mini-problems) are possible.

- An auxiliary tool is eliminated, and so is its useful action. The mini-problem is to find a yet unknown resource (X-resource) that can perform this action.
- A main tool is eliminated. The problem is similar.
- No tool was eliminated, so both useful and harmful actions are retained. The mini-problem is to find an X-resource that can eliminate or neutralize the harmful action while preserving the useful one.

- Part 3 – *Analysis of the available resources* – begins with specifying the *conflict domain*, a space within the system, occupied by the components responsible for the system conflict. Since the conflicting components not only take up space but also exist in time, the latter becomes a resource. The *operation time* is the period within which the system conflict preferably should be resolved. Next, the inventory of the available substance, field, and time resources is put together. Substance–field resources are divided into three groups.

(1) *In-system resources.* These are the resources of the *conflict domain*, the resources of the objects and tools.

(2) *Environmental resources.* These are resources of the environment specific to the conflict domain, as well as general resources that are natural for any environment, such as magnetic and gravitational fields of the Earth.

(3) *Overall system resources.* This group contains side-products (e.g., waste), pro-duced by other systems, or any cheap or free "foreign" substances and fields.

It is recommended to first attempt the problem by utilizing in-system resources and, if this approach does not work, try resources from the other groups.

• Part 4 – *Development of conceptual solutions* – as its name suggests, contains steps for resolving the system conflicts using the identified resources. First, one of the substance–field resources – X-resource – is selected for modification. Then, an *ideal final result (IFR)*, which is an ideal functioning of the selected resource in the conflict domain, is formulated. In essence, the selected resource has to perform the actions specified in the formulation of the mini-problem.

If the IFR can be readily realized, then the problem is solved. Often, however, opposing demands to the physical state of the selected resource (a physical con-tradiction) make the direct realization of the IFR impossible. Then, the separation principles are used to resolve the physical contradiction. If necessary, two physical contradictions – macro and micro – are formulated and attempted.

If a physical contradiction cannot be formulated for the selected resource, or it can be formulated but not resolved, another resource should be chosen.

What should one do if all of the above steps have not resulted in a good concept? First, one needs to double-check that the formulation of the primary function contains no more than one physical action. For example, the primary function "to handle the part (by a robot)" may refer to a combination of such distinct physical actions as "moving the part along the X-axis" and "rotating the part." Only one of these actions may be directly related to the pertinent system conflicts; mixing these actions together may muddle up the subsequent analysis. Even if two or more actions are directly causing system conflicts, each of them must be separately analyzed.

Another opportunity to continue the analysis is to "revive" the auxiliary tool elim-inated in Part 2, along with the system conflicts it caused. Resolving these system conflicts will hardly produce a breakthrough, but may still result in a noticeable improvement of the system. If no auxiliary tool was part of the conflicting interac-tion, then another intensified system conflict (usually associated with a maxi-problem) should be analyzed and resolved.

Following are the text of ARIZ and examples illustrating its use. In many cases some steps are not relevant and can be "skipped."

4.3   Text of ARIZ

Part 1. *Formulation of system conflicts*

Steps	**Actions**, Rules, and *Recommendations*
1.1 Formulate the Primary Function, PF	*Recommendation 1: Avoid professional jargon when describing functions, components, and their actions.*
1.2 Identify major components of the system, MC	**List MC of the system and of its environment.**
1.3 Select conflicting components	**Select MC associated with useful and harmful actions.** • Object (O) • Main tool(s) (MT) • Auxiliary tool(s) (AT) • Environmental element (EE) <u>Rule 1</u>: If a main tool is absent, proceed with completion of a sufield (use the Standards of Group 1). <u>Rule 2</u>: If there is more than one pair of identical conflicting components, analyze just one pair. <u>Rule 3</u>: If there is more than one AT, rank them as enabling, enhancing, correcting and measuring.
1.4 Formulate system conflicts, SC	**(a) Build system conflict diagrams.** **(b) Formulate system conflicts SC-1 and SC-2.** <u>Rule 4</u>: System conflicts are formulated by using nouns for components and verbs for actions performed by the components. <u>Rule 5</u>: System conflicts are formulated for physical entities (i.e., substances and fields) only, not for parameters. <u>Rule 6</u>: System conflicts are formulated with respect to opposing physical sates of main and/or auxiliary tools.

Proceed to Part 2, Step 2.1.

Part 2. *Analysis of the system conflicts and formulation of a mini-problem*

Steps	**Actions**, Rules, and *Recommendations*
2.1 Do the SCs include an AT?	**If yes, go to Part 2-1, Step 2.2-1.** **If no, go to Step 2.2**
2.2 Formulate intensified system conflicts, ISC	**Intensify the system conflicts.** **(a) Formulate ISCs, ISC-1 and ISC-2, by showing extreme states (actions, conditions) of the involved components.**

(cont.)

Part 2. *(cont.)*

Steps	**Actions**, Rules, and *Recommendations*
	(b) Build system conflict diagrams for ISC-1 and ISC-2.
2.3 Select an ISC	**Select one ISC for further analysis.** Rule 7: Select an ISC, which better emphasizes the PF. Rule 8: If both ISCs result in the impossibility of performing the PF of the system, or of the overall system, select an ISC (or formulate a new ISC-3) associated with an absent main tool.
2.4 Perform sufield analysis of the selected ISC	• **Build a sufield model representing the selected ISC.** • **Identify a Standard that may be used to resolve the ISC.**
2.5 Formulate a mini-problem	**Specify actions of an X-resource capable of resolving the selected ISC.** It is required to find an X-resource that would preserve (specify the useful action, UA), while eliminating or neutralizing (specify the harmful action, HA).

Proceed to Part 3, Step 3.1.

Part 2-1. *Analysis of the system conflicts and formulation of a mini-problem*

Steps	**Actions,** Rules, and *Recommendations*
2.2-1 Eliminate the AT	*Recommendation 2: The preferred order of elimination of AT's:* • *enhancing* • *correcting and measuring* • *enabling.*
2.3-1 Formulate SC-3	• **Formulate a new SC-3 according to the pattern:** "If the AT is absent, then (specify the HA) disappears, but (specify the UA, performed by the present AT) also disappears."
2.4-1 Perform a sufield analysis of SC-3	• **Build a sufield model representing SC-3.** • **Identify a Standard that may be used to resolve SC-3.**
2.5-1 Formulate a mini-problem	**Specify actions of an X-resource capable of resolving SC-3.** It is required to find an X-resource that would provide (specify the UA of the absent AT) without hindering the performance of the PF.

Proceed to Part 3, Step 3.1.

Part 3. *Analysis of the available resources*

Steps	**Actions**, Rules, and *Recommendations*
3.1 Specify the conflict domain, CD	**Define the space within which the conflict develops.**
3.2 Specify the operation time, OT	**Define the period of time within which the system conflict should be overcome.** • Pre-conflict Time T_1. • Conflict Time T_2. • Post-conflict Time T_3.
3.3 Identify the substance–field resources	**Resources of the CD.** (a) Substance(s) of the main tool. (b) Field(s) of the main tool. (c) Substance(s) of the object. (d) Field(s) of the object. **Resources of the environment.** (e) Substances in the environment. (f) Fields in the environment. **Resources of the overall system.** (g) Substances in the overall system. (h) Fields in the overall system.

Proceed to Part 4, Step 4.1.
In case of an absent AT (Steps 2.1-1–2.4-1), proceed to Part 4-1, Step 4.1-1.

Part 4. *Development of conceptual solutions*

Steps	**Actions**, Rules, and *Recommendations*
4.1 Select an X-resource for modification	Select one of the resources from Step 3.3 for modification. Rule 9: Select resources in the CD first. *Recommendation 3: The preferred sequence of resource modification.* (1) *Modification of a tool* (2) *Modification of the environment* (3) *Modification of an object in the CD* (4) *Modification of the overall system.*
4.2 Formulate an ideal final result (IFR) for the selected resource	**(a) The IFR can be formulated as follows.** The selected X-resource provides (use the UA from Step 2.5) during the OT within the CD, while eliminating or neutralizing (use the HA from Step 2.5). **(b) Examine an opportunity to realize the IFR. If the answer is positive, formulate a concept of solution; otherwise, go to Step 4.3.**
4.3 Formulate a physical contradiction macro	• **To provide the UA, the selected X-resource must have (specify a macro-property P).**

(cont.)

Part 4. *(cont.)*

Steps	**Actions,** Rules, and *Recommendations*
	• **To eliminate the HA, the selected X-resource must have property (specify an opposite macro-property –P).**
	<u>Rule 10</u>: If a physical contradiction macro cannot be formulated, return to Step 4.1 and select another resource. Repeat Steps 4.2–4.3 for that resource.
4.4 Resolve the physical contradiction macro	• **Use the separation principles individually or in combination.**
4.5 Formulate a physical contradiction micro	• **To have micro-property P, the selected resource must consist of particles with property (specify property μP).** • **To have micro-property –P, the selected resource must consist of particles with property (specify property –μP).**
4.6 Resolve the physical contradiction micro	• **Use the separation principles as at Step 4.4.**

4.7 If the mini-problem has not been solved	• **Examine the following opportunities.**
	   • Select another resource (Step 4.1) and repeat Steps 4.2 through 4.6.
	   • Check the formulation of the mini-problem at Step 2.5; it may be a cluster of two or more different problems.
	   • If, after performing the above two sub-steps, the mini-problem still has not been solved, select another ISC (i.e., solve a maxi-problem).

Part 4-1. *Development of conceptual solutions*

Steps	**Actions,** Rules, and *Recommendations*
4.1-1 Select an X-resource for modification	Select one of the resources from Step 3.3 for modification. <u>Rule 11</u>: Select resources in the CD first. *Recommendation 3: The preferred sequence of resource modification.* *(5) Modification of a tool* *(6) Modification of the environment* *(7) Modification of an object in the CD* *(8) Modification of the overall system.*
4.2-1 Formulate an ideal final result (IFR) for the selected resource	**(a) The IFR can be formulated as follows.** The selected resource provides (use the UA of the absent AT from Step 2.4-1), while not hindering performance of the PF. <div align="right">*(cont.)*</div>

Part 4-1. *(cont.)*

Steps	Actions, Rules, and *Recommendations*
	(b) Examine an opportunity to realize the IFR. If the answer is positive, then formulate a concept of solution; otherwise, go to Step 4.3.
4.3-1 Formulate a physical contradiction macro	• **To provide the UA, the selected must have (specify a macro-property P).** • **To eliminate the HA, the selected resource must have property (specify an opposite macro-property –P).** Rule 12: If a physical contradiction macro cannot be formulated, return to Step 4.1-1 and select another resource. Repeat Steps 4.2-1–4.3-1 for that resource.
4.4-1 Resolve the physical contradiction macro	• **Use the separation principles individually or in combination.**
4.5-1 Formulate a physical contradiction micro	• **To have macro-property P, the selected resource must consist of particles with property (specify property μP).** • **To have macro-property –P, the selected resource must consist of particles with property (specify property $-\mu P$).**
4.6-1 Resolve the physical contradiction micro	• **Use the separation principles, as at Step 4.4.**

4.7-1 If the mini-problem has not been solved	• **Examine the following opportunities.**
	  • Select another resource (Step 4.1-1) and repeat Steps 4.2-1 through 4.6-1.
	  • Check the formulation of the mini-problem at Step 2.4-1; it may be a cluster of two or more different problems.
	  • Return to Step 2.2, restore the AT and proceed with the analysis.

Example 4.1

Some systems using vacuum contain pipe lines made from glass pipes welded together. After the pipes are welded, the welds have to be hole-checked. To accomplish this, an electrode is inserted inside the pipe and positioned against the weld (Fig. 4.4). Another electrode is positioned outside the weld. When high electric voltage is applied to the electrodes, a corona discharge develops across the glass (this is seen as a halo between the electrodes). If there is even a tiny hole in the weld, the discharge concentrates inside of it, and the air in the hole shines very brightly. Then, the outside electrode is removed and the hole is closed with a gas welding torch. As soon as the

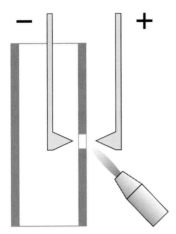

Fig. 4.4. Detecting and patching up holes in glass pipes.

outside electrode is removed, however, the hole is no longer seen, and one might be forced to melt a wider weld area than is needed (to ensure that the hole is closed). This is not desirable because it may cause harmful thermal stresses. If the hole detection coincides with the hole closing, the outside electrode is quickly burned.

Part 1: Formulation of system conflicts

1.1. **Primary function**

Two processes are taking place in the described system: detection of potential holes (or, more precisely, air) in the glass pipes and closing of the detected holes (i.e., melting the glass around the holes). The latter process is apparently dominant, so the primary function can be formulated as: *to melt the glass pipe.*

1.2. **Major components**

Glass, torch (flame), electrodes (outer and inner), air (in the hole).

1.3. **Conflicting components**

Object: Pipe
Main tool: Torch (flame)
Auxiliary tool: Outer electrode
Environmental element: Air (in the hole)

1.4. **System conflicts**

System conflict 1 (SC-1; Fig. 4.5)

If the outer electrode is present near the hole, it detects the air in the hole (i.e., ionizes the air), but is destroyed by the flame that (properly) welds the glass.

System conflict 2 (SC-2; Fig. 4.6)

An absent electrode cannot ionize the air in the hole, but is not destroyed by the flame that may overheat the glass.

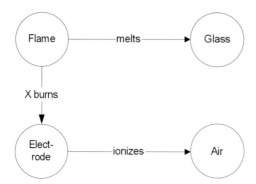

Fig. 4.5. System conflict 1.

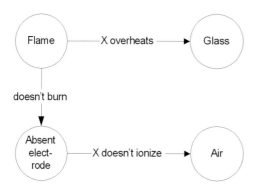

Fig. 4.6. System conflict 2.

Part 2: Analysis of the system conflicts and formulation of a mini-problem

2.1. **Checkpoint 1**

Since both SC-1 and SC-2 involve a conflicting auxiliary tool (the outer elec-trode), the analysis needs to be continued at Step 2.2-1.

2.2-1. **Elimination of the auxiliary tool**

The outer electrode – a measuring auxiliary tool – should be eliminated.

2.3-1. **Formulation of system conflict 3 (SC-3)**

Clearly, SC-3 is SC-2.

2.4-1. **Sufield analysis of SC-3**

The sufield model is shown in Fig. 4.7: In the absence of the outer electrode, an electric field is not applied to the air, which, in turn, does not generate light; this calls for adding a second substance that should be selected from the already available resource components. Obviously, this resource component has to be electroconductive to transmit the electric field to the air.

2.5-1. **Formulation of a mini-problem**

It is required to find such an X-resource that would apply the electric field to the air without hindering the welding of the glass.

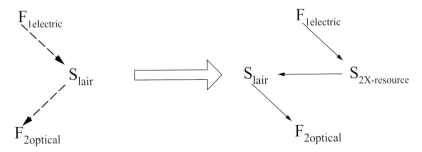

Fig. 4.7. Some existing electroconductive component should serve as an outer electrode.

Part 3: Analysis of the available resources

3.1. **Conflict domain**
This is the space formerly occupied by the outer electrode, and the vicinity of this space.

3.2. **Operation time**
This is the period before and during the welding.

3.3. **Identification of the substance–field resources**
All of the three resource groups – in-system (conflict domain), environmental, and overall – should be considered when solving real-life problems. In this chapter, for simplicity's sake, only the resources of both the conflict domain and the environment will be considered. In this case, the resources of the conflict domain are: the flame (plasma, i.e., ions and electrons), glass, air, and an electrical field. The dominant resource in the environment is the air.

Part 4: Development of conceptual solutions

4.1-1. **Selection of the X-resource**
Since the flame is the main tool, it should be chosen as the first candidate for the X-resource.

4.2-1. **Ideal final result (IFR)**
The IFR is formulated by substituting the "X-resource" in the formulation of the mini-problem (Step 2.5-1) with the name of the selected resource at the previous preceding step.

The flame in the conflict domain during the operation time applies the electric field to the air without hindering its ability to weld the glass. Can the flame satisfy these demands? The answer is: yes, since the flame (plasma) is a collection of electrically charged particles capable of conducting electricity. The flame can serve as an electrode before the welding begins (Fig. 4.8).

Example 4.2
Military turbo-fan engines often employ after-burning. During after-burning, fuel is added and ignited in the exhaust of a turbo-fan engine to create additional thrust. When

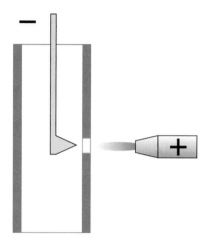

Fig. 4.8. Using the torch as an electrode.

the exhaust burns, it becomes less dense and requires a wider opening of the exhaust nozzle to maintain the air flow rate at a level required for proper engine operation. An insufficient exhaust nozzle area limits this air flow rate and creates back-pressure on the fan, which is very undesirable. To prevent a build-up of back-pressure, an adjustable exhaust area nozzle is used. When the afterburner is on, the circumference of the nozzle's throat is increased. This allows exhaust gas to more easily escape from the exhaust chamber, thus decreasing the pressure in it.

This approach has many disadvantages. The adjustable exhaust area nozzles are typically made up of large, heavy metal flaps with associated linkages and hydraulic actuators (Fig. 4.9). Consequently, such nozzles have a greater design complexity, weight, and greater manufacturing and operation costs than do fixed-geometry nozzles.

Furthermore, each moveable flap has edges and gaps between itself and adjacent structures that increase the radar visibility of the nozzle, a very disadvantageous trait for military aircraft.

The radar signature of the adjustable exhaust area nozzle prevent using after-burners in stealth aircraft. Currently, these aircraft must have fixed geometry nozzles. As a result, the stealth characteristics of the airframe come at the cost of increased vulnerability to the crew from their not having the extra thrust that after-burning provides.

To demonstrate how different formulations of the primary function might affect the course of ARIZ analysis, first let's analyze this situation from the perspective of an afterburner system manufacturer.

Version 1

Part 1: Formulation of system conflicts

1.1. **Primary function**

The primary function of the whole afterburning system is to rapidly accelerate the aircraft.

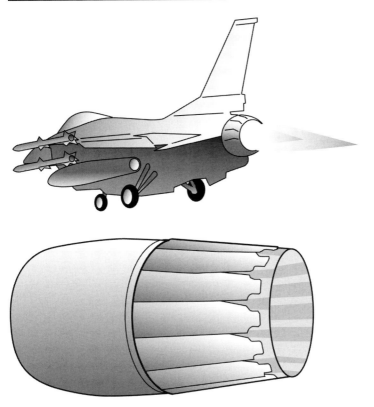

Fig. 4.9. Adjustable exhaust area nozzle with flaps.

1.2. **Main components**

Aircraft, exhaust gas, flaps, and radar signal.

1.3. **Conflicting components**

Object:	Aircraft
Main tool:	Exhaust gas
Auxiliary tool:	Flap
Environmental element:	Radar signal

1.4. **System conflicts**

System conflict 1 (SC-1, Fig. 4.10)

The flap controls (deflects) the exhaust gas (before and after the afterburning), which propels the aircraft, but it reflects the radar signal.

System conflict 2 (SC-2, Fig. 4.11)

If the flap is absent, it does not reflect the radar signal, but it also does not deflect the exhaust gas, which may adversely affect the propulsion of the aircraft.

Part 2: Analysis of the system conflicts and formulation of a mini-problem

2.1. **Checkpoint 1**

Since both SC-1 and SC-2 involve a conflicting auxiliary tool (the flap), the analysis needs to be continued at Step 2.2-1.

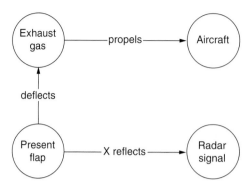

Fig. 4.10. System conflict associated with the flap.

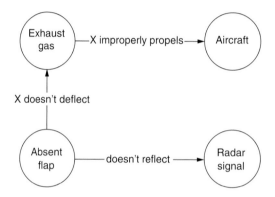

Fig. 4.11. System conflict associated with the absent flap.

Fig. 4.12. A new exhaust gas "deflector," preferably made from the existing resources, is needed.

2.2-1 **Elimination of the auxiliary tool**

The flap – a correcting auxiliary tool – should be eliminated.

2.3-1 **Formulation of system conflict 3 (SC-3)**

Clearly, SC-3 is SC-2.

2.4-1 **Sufield analysis of SC-3**

The sufield model is shown in Fig. 4.12: in the absence of the flap, the exhaust gas cannot be deflected. Both a new substance and a field have to be introduced to perform this action.

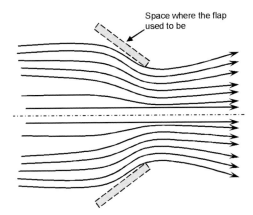

Fig. 4.13. X-resource should deflect the exhaust gas as the flap used to do.

2.5-1 Formulation of a mini-problem

It is required to find such an X-resource that would deflect the exhaust gas before and after the afterburning, without hindering the propulsion of the aircraft.

Part 3: Analysis of the available resources

3.1. Conflict domain

The space formerly occupied by the flap (Fig. 4.13).

3.2. Operation time

The periods before and during the afterburning.

3.3. Identification of the substance–field resources

The only substance resource in the conflict domain is the exhaust gas, and the available fields therein are mechanical and thermal. In the environment, the abundant substance–field resource is the cold air.

Part 4: Development of conceptual solutions

4.1-1. Selection of the X-resource

Since the conflict domain does not contain a tool, a substance from the environment (S_2 in Fig. 4.12) should be used as the X-resource. Obviously, the air is the X-resource.

4.2-1. Ideal final result (IFR)

The air in the conflict domain during the operation time deflects the exhaust gas without hindering the propulsion of the aircraft.

One conceptual embodiment of this idea is shown in Fig. 4.14. In this design, the effective flow area is varied by injecting pressurized air at selected locations along the perimeter of the nozzle. The pressurized air constricts the area available

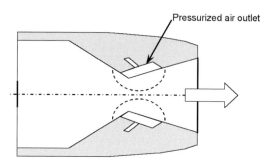

Fig. 4.14. Outside air, pumped into the nozzle, serves as a "flap."

for exhaust gas, thus aerodynamically blocking its flow area. As a result, the nozzle can be mechanically fixed in geometry, without any need for moveable flaps. The nozzle weight is low because there are no actuators or moving parts, and the structure is more efficient. The nozzle may have any desired shape and is therefore more easily integrated into the structural design of an aircraft. The surfaces of the nozzle are smooth, without any gaps, which significantly reduces radar signature.

Now let's analyze the same situation from the standpoint of a hypothetical flap manufacturer.

Version 2

Part 1: Formulation of system conflicts

1.1. **Primary function**
    The flap exists to control (deflect) the gas.
1.2. **Main components**
    Aircraft, exhaust gas, flaps, and radar signal.
1.3. **Conflicting components**

Object:	Exhaust gas
Main tool:	Flap
Auxiliary tool:	None
Environmental element:	Radar signal

1.4. **System conflicts**
    System conflict 1 (SC-1; Fig. 4.15)
        The flap deflects the exhaust gas (before and after the afterburning), but reflects the radar signal.
    System conflict 2 (SC-2; Fig. 4.16)
        If the flap is absent, it does not reflect the radar signal, but it does not deflect the exhaust gas.

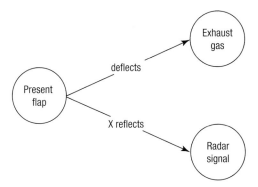

Fig. 4.15. System conflict associated with the present flap.

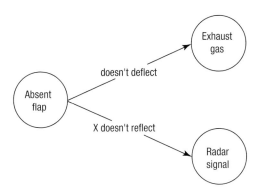

Fig. 4.16. System conflict associated with an absent flap.

Part 2: Analysis of the system conflicts and formulation of a mini-problem

2.1. **Checkpoint 1**

Since both SC-1 and SC-2 do not involve a conflicting auxiliary tool, the analysis needs to be continued at Step 2.2.

2.2. **Intensification of the system conflicts**

ISC-1 (Fig. 4.17)

The flap deflects the exhaust gas, but it reflects an amplified radar signal (which significantly increases the probability for an enemy missile to hit the aircraft).

Clearly, further intensification of SC-2 would not result in a qualitative change of the conflicting interactions (i.e., ISC-2 is effectively SC-2).

2.3. **Selection of the intensified system conflict**

Both ISCs are associated with ultimate harm to the aircraft. This triggers Rule 7 – the flap has to be eliminated – and selection of ISC-2.

From this step on, the following analysis (and its outcome) is similar to the corresponding part of Version 1.

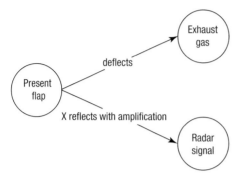

Fig. 4.17. ISC-1 is created by aggravating the harmful action of the flap.

Generally, since any technological system is part of some higher-level system, it is desirable to perform separate ARIZ analyses for two primary functions: one formulated for the given system, another for the higher-level system. If both analyses result in the same concept (as was the case in Example 4.2), this is a reliable indication that this concept is very promising. Conversely, if the results disagree, then the concept associated with the higher-level system should be given a preference. For example, if the outcomes of Versions 1 and 2 of the above ARIZ analysis had been different (say, Version 2 led to a conclusion that the flap should be modified, but still retained), then the outcome of Version 1 – eliminating the flap – should have been deemed more promising (since it is associated with a higher-level afterburning system). Such a biased treatment of the concepts follows from the notion of the evolutionary dominance of the higher-level systems (more on this in Chapter 6).

Example 4.3
Plasma cutting is a process that uses a high-velocity jet of ionized gas (i.e., plasma) to cut metal workpieces (Fig. 4.18). To create plasma, an electric arc is struck between the electrode and the workpiece. The high-temperature plasma heats the workpiece, melting the metal. The high velocity plasma jet blows the molten metal away, thus cutting the workpiece. A significant problem with conventional plasma cutting is the wear of electrodes whose temperature can reach 3000°C. Typically, the electrodes include a hafnium, or a zirconium, or a titanium insert. These materials are desired for their ability to withstand high temperatures, but are very costly and require frequent replacement. Cooling of the electrodes complicates the overall system and is only marginally effective.

Part 1: Formulation of system conflicts

1.1. **Primary function**
     To cut the workpiece.

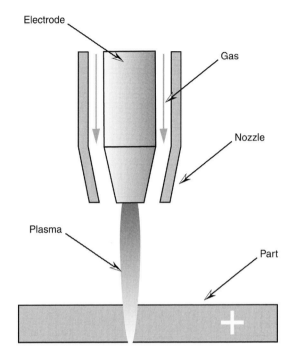

Fig. 4.18. Principal parts of a plasma torch apparatus.

1.2 **Main components**
   Plasma, workpiece, gas, nozzle, electrode.

1.3 **Conflicting components**

Object:	Workpiece
Main tool:	Plasma
Auxiliary tool:	Electrode

1.4 **Formulation of system conflicts**
   System conflict 1 (SC-1; Fig. 4.19)
      A powerful plasma cuts (melts) the workpiece quickly, but also quickly consumes (melts) the electrode.
   System conflict 2 (SC-2; Fig. 4.20)
      A less powerful plasma melts the workpiece more slowly, but also melts the electrode more slowly.

Part 2: Analysis of the system conflicts and formulation of a mini-problem

2.1. **Checkpoint 1**
   Since both of these system conflicts involve the auxiliary tool, the analysis continues at Step 2.2-1.

2.2-1. **Elimination of the auxiliary tool**
   The electrode – an enabling auxiliary tool – should be eliminated.

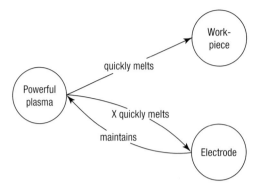

Fig. 4.19. SC-1 involves a powerful plasma.

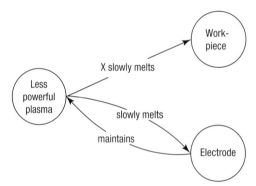

Fig. 4.20. SC-2 is due to a less powerful plasma.

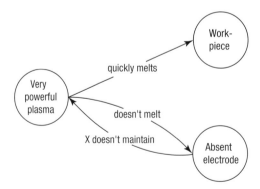

Fig. 4.21. Removal of the electrode produces a new system conflict.

2.3-1. **Formulation of system conflict 3 (SC-3)**

An "absent electrode" cannot be consumed by the plasma, but it also makes performance of the primary function impossible (Fig. 4.21).

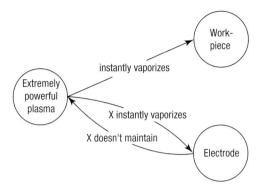

Fig. 4.22. ISC-1 is created by maximizing the plasma power.

One can now easily anticipate what should happen at the following steps of ARIZ: The electrode's useful action – supplying the electric current to the plasma (in brief, maintaining the plasma) – should be assigned to one of the available components. This component has to satisfy two critical demands: it must be both electroconductive and resistant to very high temperatures. While it is easy to identify a component made from an electroconductive material (e.g., the nozzle), the second demand can be met by no other available component. Thus, further analysis (and concept development) can follow one of three potential paths.

- Switching to, or development of, a new – plasma-less – process for cutting thick metal plates.
- Development of a new, electrode-less, process for generating powerful plasma. Such a process could employ the physical phenomena used in scientific experiments on creating high-temperature plasma. In these experiments, a very strong magnetic field, generated by superconducting coils, strips electrons off of the gas atoms, producing the plasma.
- Finding a new approach to the significant extension of the electrode's service life.

Although theoretically possible, the first two paths may be associated with extensive R&D efforts and/or significant retooling costs (the maxi-problem approach), and therefore deemed impractical by most users of the conventional plasma cutting equipment. Consequently, to continue with this case study and stay within the mini-problem approach, the electrode should be "restored," and the analysis resumed at Step 2.2.

2.2. **Intensification of the system conflicts**

ISC-1 (Fig. 4.22)

An extremely powerful plasma torch instantaneously vaporizes both the work-piece and the electrode.

ISC-2 (Fig. 4.23)

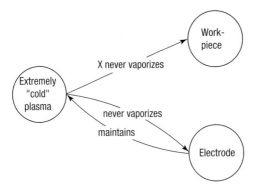

Fig. 4.23. ISC-2 is created by eliminating the plasma.

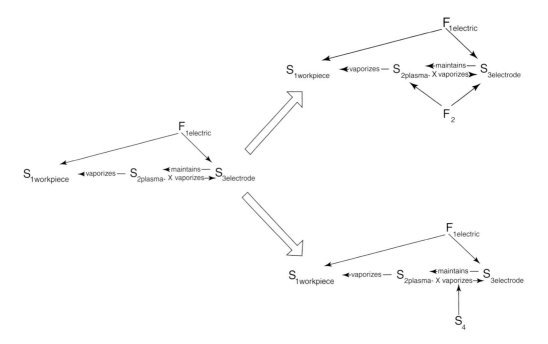

Fig. 4.24. Sufield models of potential solutions.

An extremely "cold" (i.e., absent) plasma does not vaporize both the workpiece and the electrode.

2.3. **Selection of the intensified system conflict**

Since ISC-2 is associated with an impossibility to perform the primary function, ISC-2 should be selected for further analysis.

2.4. **Sufield analysis of SC-3**

The left part of the sufield "equation" in Fig. 4.24 shows a complete sufield whose two substances participate in both useful and harmful interactions. One of the

Standards for Elimination of Harmful Actions (or a combination of these) described in Chapter 3, can be used to prevent (or at least to dramatically retard) the vaporization of the electrode.

2.5. Formulation of a mini-problem

It is required to find such an X-resource that would prevent the vaporization of the electrode by the plasma without hindering the vaporization of the part.

Part 3: Analysis of the available resources

3.1. Conflict domain

The conflict domain includes the parts of the conflicting components that directly create the useful and harmful actions: the tip of the electrode, the whole plasma and the section of part affected by the torch.

3.2. Operation time

The part cutting period.

3.3. Identification of the substance–field resources

The substance resources in the conflict domain are the electrode and the plasma. The readily available fields therein are electric and thermal.

In the conflict domain's environment, the resource substances are the gas (not yet converted into plasma), the nozzle and the air. The field resource is the mechanical field and the moving gas.

Part 4: Development of conceptual solutions

4.1. Selection of the X-resource

Recommendation 3 of ARIZ helps select the plasma (main tool) for modification.

4.2. Formulation of ideal final result (IFR)

The plasma in the conflict domain during the operation time neutralizes the vaporization of the electrode, without hindering the vaporization of the part. Obviously, the plasma as is cannot accomplish this IFR.

4.3. Formulation of a macro physical contradiction

- To prevent the electrode's vaporization of the electrode, the plasma must not come in contact with it.
- To maintain the vaporization of the part (i.e., the primary function), the plasma must contact the electrode.

4.4. Resolving the macro physical contradiction

Let's use the separation principles described in Chapter 2.

Separating the opposing demands in time means that the plasma contacts the electrode for only a short period, insufficient to vaporize its material. This results in a sharp reduction of the production rate, and is, therefore, unacceptable.

Separation of the opposing demands in space means that the plasma contacts only one part of the electrode. This also does not make sense.

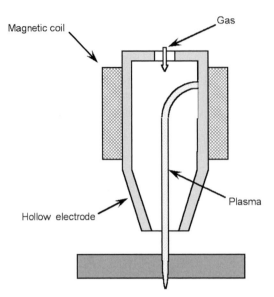

Magnetic coil

Gas

Hollow electrode

Plasma

Fig. 4.25. Moving plasma significantly expands the electrode's service life.

Separation between the whole and its parts suggests that the whole plasma does not contact the electrode, while some of its components do. Envision an embodiment of this approach is difficult.

Since applying the separation principles individually did not result in a good concept, some combinations of these principles should be attempted. For example, separation of the opposing demands in both time *and* in space. This implies that for a very short moment, the plasma contacts just one spot on the electrode's surface; the next moment, it moves to another spot. Since heat transfer between the plasma and the electrode cannot happen instantaneously, rapidly shifting the plasma across the electrode's surface can significantly slow down the erosion of the electrode. The sufield model in Fig. 4.24 suggests that a new field, F_2, should be introduced to control the plasma movement. Simple analysis of the available resource fields that can accomplish this task points to electric and magnetic fields as well as the mechanical field of the moving gas.

The sketch in Fig. 4.25 shows a long-lasting electrode based on this principle (the detailed description is in US Patent No. 4,034,250). The gas is supplied to the workpiece through a cavity in the electrode. The upper root of the plasma arc travels inside the hollow electrode under the action of the vortical air flow and rotates intensively about its inner surface. The magnetic field produced by the coil enveloping the electrode creates the electrodynamic force that also contributes to moving the upper root of the plasma arc. This force varies proportionally to the operating current.

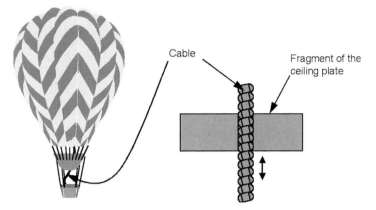

Fig. 4.26. Cable passes through the ceiling plate to control the outlet valve in the gondola.

4.4   Summary

ARIZ is the central analytical tool of TRIZ. Its basis is a sequence of logical procedures for the analysis of a vaguely or ill-defined initial problem/situation, and transforming it into a distinct system conflict. Analysis of the system conflict leads to the formulation of a physical contradiction whose elimination is provided by the maximal utilization of the resources of the subject system. ARIZ brings together most of the fundamental concepts and methods of TRIZ such as ideal technological system, system conflict, physical contradiction, sufield analysis, the Standards, and the laws of technological system evolution.

4.5   Exercises

4.1 Many heat exchange devices use fluids (liquids or gases) that flow parallel to the hot surface. As the fluid flows along the surface, it becomes hotter and therefore less capable of taking heat because the temperature difference between the fluid and the surface decreases. Develop a better method for heat transfer between a surface and a fluid.

4.2 On May 27, 1931, Auguste Piccard with his associate Paul Kipfer, made the first manned balloon flight into the stratosphere, ascending to a then-record altitude of 51 961 feet (15 838 m).

Piccard's stratospheric balloon consisted of a gas balloon and an airtight pressurized passenger gondola made of aluminum (Fig. 4.26). A control cable for the outlet valve of the balloon had to pass through the ceiling of the gondola. In order to

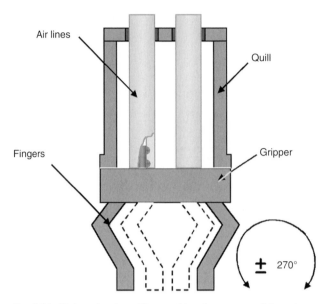

Fig. 4.27. Nylon air tubes affect positional accuracy of the gripper.

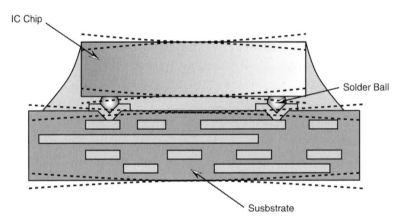

Fig. 4.28. Conventional mounting of IC chips.

Fig. 4.29. Worm drive.

keep the air inside the gondola, the gap between the cable and the ceiling plate should not have existed. But this would have prevented free motion of the cable. Various tested sealing devices were not reliable. Use ARIZ to reinvent the solution found by Piccard.

4.3 A robot gripper is powered by two air lines made from hard nylon robes (Fig. 4.27). One air line powers the gripper fingers, the other powers a pneumatic motor that rotates the gripper. The lines are firmly attached to the top of the quill and to the gripper, which rotates $\pm 270°$. Rotation of the gripper causes a twisting of the nylon tubes, which results in a high and varying torsional resistance to the gripper motion and, consequently, in less positioning accuracy.

Softer PVC tubes exhibit fast fatigue failures at their attachments to the grippers. How can a low torsional resistance be achieved while using hard nylon tubes?

4.4 Usually IC chips are mounted on supporting substrates made of a material with a coefficient of thermal expansion that is different from that of the material of the IC. This causes thermal expansion stresses that may crack interconnects (Fig. 4.28). Find ways to alleviate the ill effects of the different thermal expansion rates.

4.5 It is known that the efficiency of a worm drive (Fig. 4.29) is determined by friction coefficient in the contact area and by the lead angle of the worm. The efficiency can be improved by increasing the lead angle. However, increasing the lead angle reduces the transmission ratio of the worm drive for the given worm/wheel diameters. Develop concepts of a worm drive that can satisfy these conflicting demands.

5 Laws of technological system evolution

The concept of predictable evolution of technological systems, proposed by Genrikh Altshuller, implies that various contributing factors such as the efforts of individuals producing breakthrough inventions, market forces, political conditions, cultural traditions, etc. may greatly affect the pace of this evolution (speed it up or slow it down), but they cannot significantly alter its direction. The steam engine, electric motor, automobile, airplane, radio, ball pen, laser, transistor, integrated circuit, and all other major innovations would have been invented sooner or later even if their actual creators were not born.

This chapter describes the laws of technological system evolution that are the conceptual foundation of TRIZ. A judicious use of these laws may greatly contribute to success in business and engineering practices, since technology leaders and designers are able to identify the most promising innovations among numerous candidates. These laws can also be used to efficiently develop novel technologies and products, to objectively assess those systems' business potential, and to predict what competitors may come up with.

5.1   Laws of technological system evolution

Technological systems are not frozen entities. They are in a state of continuous change in order to satisfy the changing needs of society and to survive in a very competitive global market. The evolution of any technological system can be presented as a set of stages, with the system at each stage being distinct from the system at the previous stage. Between these stages, the system may undergo numerous developments which gradually accumulate and then bring the system to the next stage.

The dynamics of the evolution of technological systems can be illustrated by the development of automobile designs over the last 50 years. Designs vary from huge structures with fins in the 1950s, to sleek modern cars that look like midgets compared to their predecessors, but often have a comparable or even larger interior space. Hundreds of car models have incorporated various design changes dictated by aerodynamic requirements, the availability of new materials and systems (e.g., air bags, anti-lock brakes), changes in powertrain (e.g., front wheel drive, smaller but powerful engines), customer tastes and, of course, the artistic tastes of designers. Many changes have disappeared without a trace, others have evolved further, but every 3 to 5 years a new design trend becomes obvious.

Technological systems are constantly under pressure from society, directly or through market forces, to evolve into better performing systems, into systems requiring less resources to manufacture, or into systems that combine several performance functions. This evolution is represented by incremental changes as well as by breakthroughs. These changes usually occur due to inventions. Thousands of people, not only qualified engineers, make big and small inventions every year. Only a small fraction of the inventions are implemented, but all of them contribute, both to the state of the art of engineering in general and to the state of the art in their particular area (i.e., for a specific class of technological systems).

The pool of inventors is not regulated. Inventors are largely independent in their creativity (albeit pressured by market and societal forces and/or by their employers), and are frequently not aware of the efforts of other inventors working with the same technological systems. Their efforts are aimed in seemingly random directions. The rejection of many inventions can be explained, to a certain extent, by such factors as a lack of capital, the psychological inertia of investors, "not invented here" attitudes, etc. In the long run, however, the ultimate acceptance of an invention depends on its compliance with the logic of technological systems evolution.

A law of technological system evolution describes significant, stable, and repeatable interactions between elements of the system, and between the system and its environment in the process of its evolution.

The known laws of evolution have already been listed in Chapter 1, but it is worth listing them again here:
- law of increasing degree of ideality
- law of non-uniform evolution of sub-systems
- law of transition to a higher-level system
- law of increasing dynamism (flexibility)
- law of transition to micro-level
- law of completeness
- law of shortening of energy flow path
- law of increasing substance–field interactions
- law of harmonization of rhythms.

These laws were formulated after the analysis of tens of thousands of patents and the historical development of numerous real-life systems. This process included the selection of breakthrough inventions, and their examination, so that typical phases in the systems' evolution could be discerned.

Not all of these laws are universally applicable to all stages of systems' evolution. Some of them – the laws of increasing ideality, non-uniform evolution of sub-systems, transition to a higher-level system, and increasing flexibility – govern the evolution of technological systems at all stages of their historical development. For example, all technological systems, whether they are newly developed or matured, experience an uneven rate of evolution of their different subsystems. This non-uniform evolution results in conflicts among the subsystems that need to be resolved.

The other laws determine the systems' evolution only at certain stages. The law of harmonization of rhythms is an example. The principal parts of the most viable systems have their functions synchronized (harmonized) with each other. Once such harmonization of rhythms has been achieved, this law is no longer applicable.

Unlike the laws of nature, the laws of evolution are "soft," and can be temporarily "transgressed." While it is impossible to violate or ignore any law of nature, violations of the laws of evolution are very common. Quite often, various non-technological factors, such as an economic pressure to continue the utilization of expensive existing manufacturing facilities, political interests, psychological inertia and the like may cause a delay in the emergence of next-generation products or processes. Deviations from the path of evolution usually results in a waste of always precious resources and lost competitiveness.

Example 5.1

The fall of the Swiss watch industry is enlightening in this regard. In 1974, Swiss mechanical watch manufacturers enjoyed close to 40% of the world export market (Glasmeier, 1997). Electronic watches were introduced at about that time. The transition from mechanical watches to electronic ones should have occurred sooner or later because the migration from mechanical systems to systems based on electric phenomena is "prescribed" by the laws of transition to micro-level and increasing flexibility. The Swiss watch industry leaders were skeptical about the viability of the radically new timekeeping devices. Ten years later, the share of Swiss mechanical watches in the world market fell precipitously to just 10% as a result of the advent of electronic watches pioneered by Japanese watch makers.

5.1.1   Lines of evolution

Any law of evolution defines a general direction of the development for next-generation systems. The predictive power of some of these laws is significantly reinforced by associated *lines of evolution* that specify stages of a system's evolution along a general direction. A law of evolution relates to its lines of evolution just as a compass relates to a roadmap: the former helps to determine the direction to the destination, while the latter shows particular paths that one may travel to reach this destination. For example, the law of increasing flexibility states that in the course of evolution, technological systems become more dynamic, more adaptable. Lines of increasing flexibility identify certain flexibility phases, e.g., "segmented system," "elastomeric system," etc.

Not all systems advance through all the stages; depending on the nature of a particular system, it may skip over some stages on a specific line of evolution.

5.2   Law of increasing degree of ideality

This is the primary law of the evolution of technological systems.

Evolution of technological systems proceeds in the direction of an increasing degree of ideality.[1]

This means that in the process of evolution, either a system performing certain functions becomes less complicated, costly and/or "problematic," or it becomes capable of performing its functions better, or it performs more functions. A combination of these evolutionary processes also often occurs.

Example 5.2

Figure 5.1 shows the dynamics of prices for household appliances in the US from 1947 through 1997 (in 1997 US dollars). Total functionalities of 1997 washers/dryers, TV sets, and refrigerators are noticeably higher, while their prices are much lower than those of their 1947 predecessors. The comparison clearly shows how appliances have come a long way towards more ideal systems.

An increasing degree of ideality can be seen in the evolution of many familiar systems.

Prices of cellular telephones have dropped almost tenfold over the last decade, and so did their power consumption. In this same period, cellular phones acquired new capabilities: e-mail, GPS, PDA, TV, games and others (some extra functions are probably being introduced even as this book is being written).

Although prices of modern high-power/high-accuracy/high-speed machine tools are increasing, the cost of a part produced on these machine tools is decreasing due to their continuously growing productivity. At the same time, the accuracy and surface finish of the produced parts are continuously improving.

Systems with a higher degree of ideality have much better chances to become winners in the market selection process and to dominate the market for the long-run. This sheds some light on the process of technological choice: why do some technologies become prevailing, while others become marginalized, or even fall into oblivion?

Example 5.3

At the dawn of the twentieth century, three powerplants competed for the place under the car's hood: an internal-combustion engine, a steam engine, and an electric motor. It appears that the internal-combustion engine won this contest, since it was not only technically superior, but also had fewer non-technical problems. The thermal efficiency

---

[1] The concept of degree of ideality was introduced in Chapter 2.

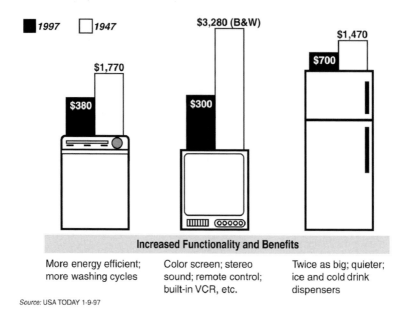

Source: USA TODAY 1-9-97

Fig. 5.1. Dynamics of degree of ideality for household appliances.

of a then-nascent gasoline engine was higher than that of a highly developed steamer. It also allowed for longer periods between "recharging" than a battery-powered electric motor. These technical advantages, however, were not the only factors that determined the eventual triumph of the gasoline engine.

In 1913, a British observer noted that "the American buyer never inquires about mechanism. He wishes to know merely what price you want for a car . . . He has paid for his car, and he is going to do what he pleases with it without any feeling for mechanics at all; whereas the average European buyer knows, or, at least, likes to think he knows, a good deal about the mechanical details of his chassis" . . . The legendary Stanley Steamer was festooned with gauges that required regular attention: boiler water level, steam pressure, main tank fuel pressure, pilot tank fuel pressure, oil sight glass, and tank water level. Just to get it started required the manipulation of thirteen valves, levers, handles, and pumps. The electric car served admirably as an around-town runabout, but most drivers wanted some dash and excitement in their motoring, even when their cars were largely employed for mundane purposes. The electric also suffered from the fatal defect of being identified as a "woman's car," a major failing at a time when the automobile was tightly intertwined with masculine culture" (Volti, 1996).

The law of increasing degree of ideality is cardinal among the laws of evolution, because it sets up the primary vector for the historical development of technological systems. Other laws of evolution can be regarded as mechanisms for the development of next-generation systems with an increased degree of ideality.

5.3   Law of non-uniform evolution of subsystems

Any technological system satisfies some needs which usually grow faster than the improvements of the system. This also creates an evolutionary pressure causing the development of system conflicts.

Law of non-uniform evolution of technological systems
> The rate of evolution of various parts of a system is not uniform; the more complex the system is, the more non-uniform the evolution of its parts.

This non-uniformity begets system conflicts whose resolution requires the development of new inventions, and thus promotes the evolutionary process. This law can be illustrated by the evolutions of numerous technological systems.

Example 5.4

Evolution of the bicycle
Table 5.1 shows the milestones of bicycle evolution (Altshuller and Shapiro, 1956). It is important to remember that Table 5.1 reflects the evolution of only the primary system; there are many smaller branches on the evolution tree, e.g., the development of practical transportable bikes (foldable or disassemblable), etc.

Example 5.5

A computing system consists of a computer and peripheral devices. One important peripheral device is a printer. In the 1960s, teletypes (which were invented in the 1930s for sending telegrams) were satisfying the needs of relatively slow computers. Computing speeds were increasing at a fast pace, however, especially after the perfection of microchip technology. The high computational speed was, in some cases, useless since it could not be matched by teletypes using comparatively heavy mechanical moving parts. Also, high-speed computers could handle graphic problems, but only plotters could provide the hard-copy graphics. This system conflict led to a massive effort towards the development of output devices which could provide for high-speed printing and make a hard copy of any pattern (various fonts and alphabets, combinations of alphanumeric and graphical information, desktop publishing). These developments followed the following path.
- Fast alphanumeric printing with the possibility of the fast changing of character carriers (fast printers with character balls and daisy wheels); they provided for a limited flexibility.
- Introduction of universal light printing elements (pins) which enabled the generation of any character or pattern; such dot printers had much greater flexibility (see Section 5.4 below), but rather low print quality since the size of the dots and

Table 5.1. *Milestones of bicycle evolution*

YEAR	BICYCLE GENERATION	SYSTEM CONFLICT
1813 	First bicycle: frame, two wooden wheels, and handle bar. No pedals and transmission. Rider propelled himself by pushing with his feet.	Increasing speed is limited by natural capabilities of the riders.
1840 	*Pedals* are introduced. As the wheels and pedals improved, *bicycle speed increased.* Brakes are absent, the biker's shoe soles perform braking.	Increased speed endangered the rider, *since the braking function becomes inadequate.*
1845	*Brakes introduced.* Higher speed is achieved by increasing diameter of the front wheel. Distance traveled per pedal revolution increased.	Further speed escalation is constrained by *low durability of the material of the wheels.*
1860s 	Wooden wheels are replaced with *metal wheels.* Front wheel diameter increases, speed also increases.	The greater the wheel diameter, the more *strenuous effort is necessary to rotate the wheel.*
1870s	*Bearings* in the wheel axle are introduced. Front wheel diameter continues to increase. Accordingly, the bicycle height increases too.	The more height, i.e., wheel diameter, the higher the speed, but the *more danger for the rider* (falls from a bicycle frequently caused serious injuries).
1884 	*Chain transmission* is introduced which *eliminated the need for increased wheel diameter.* While inventors concentrated mostly on designing novel transmissions, the wheels did not undergo changes.	Bicycles *cannot be driven at high speed on rough roads.*
1890 	*Pneumatic tires* are mounted on the wheels. Speed continues to increase.	The state-of-the-art of the transmission lags behind the state-of-the-art of the wheels: *After accelerating, the bicyclist cannot control the pedals, which spin at high speed.*
1897	*The overrunning (free-wheeling) clutch* is invented; now the pedals can be kept immovable after reaching the desired speed. Since then, the bicycle has retained its familiar appearance.	

their closeness were restricted by mechanical design limitations (another system conflict).

• Ink-jet printers, which provide high printing quality at a reasonable cost, but whose relatively low resolution limits the application of these systems for complex pattern-printing tasks.

• Laser printers, which provide the best printing quality, but are also (for now) the most expensive.

Another limiting peripheral system is the information storage system. The system conflict between the fast increasing computing speed and the lagging intermediate-range memory led to continuous improvements in magnetic disk memory systems.

Example 5.6

The first refrigerator was invented by T. Moor in 1803. He was delivering butter to customers in Washington, DC, and the need for refrigeration in summer-time Washington was very acute. Moor was using a large double-wall box, with ice filling the space between the walls. *The system conflict:* the useful function was achieved, but ice was produced in the winter time, while refrigeration was needed in the summer.

Ice handling was a very difficult task. The refrigerating compressor was invented in 1850 by John Gorrie in Apalachicola, Florida. Gorrie was treating malaria patients by cooling them, but ice was being delivered from Massachusetts at a very high price ($1.25/lb). He patented the compressor but was unable to find investors due to general disbelief. The New York Times called Gorrie a crank, ridiculed his machine and said that only the Almighty could make ice. In 1890, ice-making machines went into production due to a shortage of natural ice and were initially used to produce ice for food warehouses, chocolate candy factories, etc. *The system conflict:* the first designs were very large and thus not suitable for home use.

By the end of the nineteenth century the first ice-making machines for home use were developed. *The system conflict:* these machines were guzzling fuel – wood, coal, kerosene – so that they were very expensive for an average household to run.

In 1918, General Electric started production of modern type automatic refrigerators, or 'Kelvinators,' for kitchens. *The system conflict:* the belt-driven compressor was very noisy and leaked gases (ammonium and sulfur oxide). In 1926 the compressor was surrounded by a sealed cover, reducing noise and unpleasant odors. The kitchen refrigerator without a compressor (absorption type) was invented in 1922, and the two systems have competed ever since. *The system conflict:* while absorption fridges are less noisy, they have a lower cold output.

Another system – the solid state thermo-electric refrigerator – was invented in 1957 at NASA. While refrigerators with compressors are still the leaders for home use primarily because companies continuously add new useful functions (an ice dispenser, self-cleaning freezers, etc.), there are now affordable thermo-electric coolers/heaters on the market. A matchbook-size governing module delivers the cooling power of a 10-pound block of ice. Developments in fiberglass thermal insulation materials, semiconductor cooling technology and phase-change energy storage materials are incorporated in super-efficient 25 and 27 cubic foot home refrigerators.

5.3.1   Use of the law of non-uniform evolution of subsystems

The following steps are recommended.
(1) Identify the major subsystems of the system of interest and their primary functions.
(2) Select one of the major subsystems and (mentally) considerably intensify its primary function.
(3) Identify inadequacies or negative effects in other subsystems caused by the above function enhancement. Formulate resulting system conflicts.
(4) Repeat steps 1–3 for all other major subsystems.

Example 5.7
Both the law of non-uniform evolution of systems and the above procedure can be illustrated by a strategic management problem in agriculture. The grain-growing system consists of several sub-systems: growing grain, harvesting, and handling the harvested product. When the yields were low, the grain-growing subsystem was a weak link; harvesting was not a big problem. After the yields increased dramatically due to improvements in fertilizers and in soil preparation techniques, harvesting became a bottleneck. This was resolved by the invention of the combine. Presently, combines have a much higher productivity than the handling system (trucks, local elevators) can accommodate (Fig. 5.2). The faster the combine moves, the faster the accompanying hoppers are filled, and the more of them are needed.

Amazingly, the industry is set on making even more productive combines while the handling system – the real bottleneck – is neglected. By realizing the system conflict – the higher the combine's productivity, the less adequate the harvest handling system becomes – efforts of the industry would be beneficially redirected. The company that is the first to address this issue would gain a competitive advantage.

Thus, analysis using the law of non-uniform evolution allowed for marking the direction for further development (i.e., technology forecasting, see Chapter 6), but it did not say how to realize this next step of evolution.

5.4   Law of increasing dynamism (flexibility)

A new technological system is often developed to solve a specific problem, i.e., to perform one function. This system performs in a particular environment, at specified regimes, etc. Its design reflects, accordingly, the specific needs the system has to

Fig. 5.2. Interaction between the crop harvesting and harvest handling systems.

satisfy. The system usually features rigidly defined connections between its constitutive parts which may prevent it from adapting to the changing environment. Such a system demonstrates the feasibility of the main design concept. It satisfactorily performs the primary task for which it was developed, but its application environments, as well as performance potential, are limited. Studies of the evolution of numerous systems have demonstrated that a typical process of evolution involves *dynamization (flexibility) phases*, during which the structure of the system becomes more adaptable to the changing environment and, in many cases, also becomes multi-functional.

Law of increasing dynamism

> Technological systems evolve in the direction to more flexible structures capable of adaptation to varying performance regimes, changing environmental conditions, and of multifunctionality.

This universal trend can be easily recognized in many commonly used systems.

Example 5.8

The long history of bedding is a tale of increasing adaptability. In the beginning, our ancestors slumbered on the hard ground. Then came furs and tree branches, and later still there were mattresses made by stitching hides or fabrics together and filling them with such cushioning materials as straw, animal hair, or feathers. But those mattresses were not very accommodating. Then (relatively recently) steel coil springs were introduced. The number of springs has gradually increased, thus enhancing adaptation to our curvaceous bodies. This promoted a better weight distribution and comfort. The next major steps toward enhanced softness and conformity were a waterbed and an air mattress, both perfectly adapting to the contour of the human body. Today's advanced mattresses allow for complete body impression, firmness control in different areas of the mattress, head or leg elevation, massage, etc.

Fig. 5.3. Line of transition to continuously variable systems.

Example 5.9

The first airplanes had rigid landing gears that created additional aerodynamic drag. For low flying speeds this drag was tolerated, but with increasing speeds the landing gear system became one of the least advanced subsystems of the airplane. Then it became adaptable: the landing gear is stowed during the flight and deployed only for take-off and landing.

5.4.1   Lines of increasing dynamism (flexibility)

Several lines of increasing dynamism (flexibility) have been identified thus far. These lines can be divided into two groups: lines of increasing functional flexibility of functional components of the system and lines of increasing flexibility of physical structures. For example, a mattress is a functional component whose increasing adaptability has been made possible by transition to ever more flexible physical structures – culminating in air mattresses.

Lines of increasing functional flexibility of functional system components

• *Transition to continuously variable systems*
This is the primary line of increasing flexibility. All other lines of increasing flexibility are various manifestations of this line (Fig. 5.3).

Example 5.10

The first cars had rigid transmissions between the engine and driving wheels, with the car speed adjusted by varying the engine speed. Introduction of the multi-speed gearbox allowed for an easy change of speed in accordance with environmental conditions, with a much lesser role for the engine speed change. This improved the overall performance of the car and led to an increase in the life and fuel efficiency of engines, since the optimal speed range of the internal combustion engine is rather narrow. The optimized transmission ratio helps the engine work in or near its optimal range.

Further development of the adaptability of car transmissions proceeds in several directions. One direction is increasing the number of gear shifts in manual transmissions. While 12–20 speed transmissions are used in commercial trucks, driven by professional drivers, they would be too expensive and inconvenient in cars. Another direction is to use automatic transmission with a constantly increasing number of gear shifts. The disadvantages of this are still the discreteness of the gear shifts, high friction losses in the fluid clutch and a high complexity and cost. More flexibility, with no additional effort

Fig. 5.4. Line of transition to active adaptive systems.

from the driver, is offered by a continuously variable transmission (CVT) providing optimal efficiency. Such transmissions are used in snowmobiles and are gradually coming to passenger cars, especially hybrids. Meanwhile, the most widely used is the automatic transmission. Although it has the same (or even smaller) number of transmission ratios than the manual one, it always changes the ratio at the correct moment, does not depend on the skills or mood of the driver, and provides for more comfortable driving conditions.

• *Transition to active adaptive systems*
The three main types of adaptive systems: are *passive, operator-controlled,* and *active systems* (Fig. 5.4).

Passive systems provide an adaptation of the system to the changing environment without any special power-driven or servo-controlled mechanism. A differential in the car driveline is an example of a passive system. It automatically changes the speeds of the driving wheels as required by the turning radius of the vehicle (see *Example 5.11* below).

Operator-controlled systems are capable of changing their configurations in accordance with the operator's decision. Such systems can be either operator-driven (like in a manual gearbox) or power-driven on the operator's command (like the airplane landing gear).

Active adaptation systems have sensors monitoring the environmental conditions and communicating with control means energizing actuators' which change configuration of the system (as in ABS brake systems).

While the passive adaptation systems are frequently the simplest in use and the most cost-effective and reliable, they usually require a very creative approach to develop. Active systems, on the other hand, are the most universal, especially with the proliferation of microprocessor-based computer control systems. They, however, consist of several interacting units (sensors, control blocks, actuators, power sources), and thus are usually more expensive and less reliable.

It can be stated that while active, especially microprocessor-based, adaptive systems are fast developing, the resources of the passive adaptive systems are far from being exhausted. If properly designed, they greatly enhance the performance characteristics of technological systems with relatively minor expenses (thus increasing their degree of ideality). It is also much more effective, and less costly, to use active control on a well-designed and refined passive system.

Example 5.11

The first cars had rigid rear axles that did not allow for sharp turns during which the driving outboard wheel was traveling a much greater distance than the driving inboard wheel. If the axle driven by the engine is rigid, this causes intense slippage between the inboard wheel and the ground. While it can be tolerated for slow-moving vehicles, for fast-moving vehicles this results in increased loading of the engine, fast wear of the tires, squealing noise and potential instability of the vehicle. These problems were alleviated by introducing a differential between the halves of the driving axle that automatically change the velocity ratio between the wheels in accordance with the driving conditions. Such dynamization allowed the driving wheels to adapt to the desired turning radius since the ratio of their rotational velocities was automatically adjusted by the differential as needed.

The adaptability of the conventional differential is adequate to accommodate different velocities of the non-steered wheels moving along a curvilinear path. Since the same torque is applied to both wheels, however, different frictional resistances from the ground to each wheel (e.g., puddle of water or patch of ice under one wheel) may create problems: the vehicle may get stuck since only the wheel experiencing low resistance is driven (spinning) while the other wheel becomes stationary. One of the recent development is a differential using a friction-based transmission to both wheels in which a lubricant has its friction coefficient dependent on *velocity* (such differentials are used in some Subaru cars). Thus, the torque transmitted to the fast-spinning wheels is up to five times lower than the torque transmitted to the slower rotating wheel. Such "smart" passive adaptability greatly improves the performances of cars with differentials.

The next step in this evolutionary process is the introduction of an active four-wheel steering system in which not only wheel speeds are changing in accordance to the turning radius, but the vertical planes of all four wheels are turning by the servo-controlled system in line with the turning curve, thus further reducing slippage between the wheels and the ground and allowing much smaller turning radii.

• *Transition to self-adapting systems*

In a modular system, components can be separated and recombined (Fig. 5.5). Many systems evolve toward increasing modularity. Components that were originally closely integrated become loosely coupled, thus allowing much greater flexibility for the whole system. For example, some household stoves have removable burners that can be replaced by plug-in cooking devices, such as barbecue grills and pancake griddles.

Example 5.12

The first personal computers (such as the Apple II or the Commodore PET) were rigid sets of CPUs, memory and control chips, but they rapidly evolved into modular systems enabling different combinations of components from different vendors.

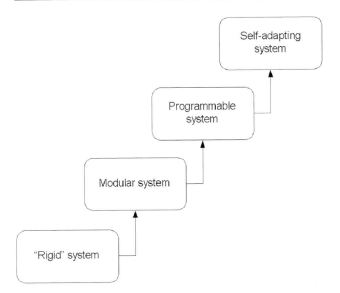

Fig. 5.5. Line of transition to self-adapting systems.

Modularity may increase at many levels of the system's hierarchy, including the level of subsystems components. In the 1980s, the development of application-specific integrated circuits (ASIC) introduced the concept of modularity into the integrated circuit chip design. As the name implies, an ASIC is a chip designed for a particular application. ASIC's are built by custom-linking existing circuit blocks (modules). Obviously, it is much easier to produce a new ASIC than to design a new chip from scratch.

The components of integrated circuits – conventional or ASIC – are tightly wired. Their final design is in the hands of a device vendor. A more adaptive approach to computer system design was introduced in the 1990s in a family of components known as "field-programmable devices" (FPD's). An FPD is an integrated circuit that a user can customize in the field to perform a task (or a set of functions). It allows system designers to program and test the circuit themselves, without involving the vendors. It can be reprogrammed an unlimited number of times. Mistakes can be corrected within hours, thus reducing testing and simulation efforts (Brown and Rose, 1996).

The most common FPD – field-programmable gate array (FPGA) – has three major configurable elements: input/output blocks (IOB), configurable logic blocks (CLB), and interconnects. The CLB's provide the functional elements for constructing a user's logic. The IOB's provide the interface between the package pins and internal signal lines. The programmable interconnects link the inputs and outputs of the CLB's and IOB's to the appropriate networks. A customized configuration is established by programming internal memory cells that determine the logic functions and internal connections implemented in the FPGA.

A radical departure from conventional chip architectures is realized in so-called "Raw chips" developed at MIT (Taylor *et al.*, 2002). These chips can *automatically*

Fig. 5.6. Line of transition toward forward sensing.

customize their internal wiring (i.e., self-adapt) *themselves* to any particular application. The software compiler takes applications written in high-level languages, such as C++ or Java, and maps them directly into the chips.

• *Transition toward forward sensing*
Transition to adaptive systems requires gauging the system's response to varying inputs and disturbances, comparing this response to the desired one and generating corrections if necessary (Fig. 5.6). Feedback is essential to guarantee good performance in the presence of uncertainty. The best performing systems can anticipate impending changes and adjust their parameters accordingly.

Example 5.13
Conventional – feedback-free – automotive suspensions use shock absorbers, springs or elastic media (rubber, compressed gas) to dampen the road-induced forces. These suspensions cannot adjust their parameters to changing road and motion conditions.

In semi-active suspensions, a microprocessor controls the orifice size of the restrictor valve in a shock absorber, thus effectively changing its spring rate. Control inputs may be vehicle speed, load, acceleration, lateral force, or a driver's individual preference.

In the most advanced active car suspensions (e.g., in the Cadillac STS continuously variable road-sensing suspension), the shock absorber settings are fine-tuned in accordance with anticipated road conditions. For example, when a front wheel hits a pothole, a signal is sent to a microprocessor that almost instantly adjusts the rear shock absorber settings before the rear wheel meets the pothole.

Lines of increasing flexibility of physical structures
• *Transition to fluids and fields*
The process of increasing flexibility often involves the replacement of stationary components with moving ones: a rigid linkage or a structure with a segmented linkage/structure connected by hinges, the replacement of rigid components with flexible ones, with hydraulic and pneumatic systems, etc., and finally – the transition to electromagnetic fields (magnetic and electric; Fig. 5.7). These fields occupy the highest level of evolution because they can be controlled within very intricate specifications, varying controlled parameters, across a three-dimensional space, etc.

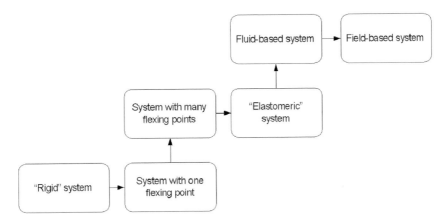

Fig. 5.7. Line of transition to increasingly flexible physical structures.

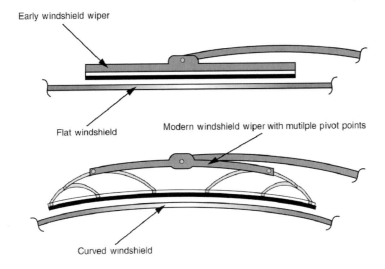

Fig. 5.8. Evolution of windshield wipers.

Example 5.14

The windshields of early cars were made of flat glass. For that reason, the first windshield wipers were straight metal bars trimmed with rubber wiper blades (Fig. 5.8). When windshields evolved into three-dimensionally shaped glass to accommodate aerody-namic and aesthetic requirements (this development itself represents a dynamization of the system), the metal bar of the wiper was modified to approximately conform to the glass shape. Small variations of the glass curvature during the wiper's travel was accommodated by the flexibility of the rubber blade. Any changes of the windshield's shape from model to model required different designs of the windshield wiper; deviation

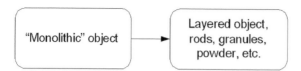

Fig. 5.9. Line of increasing fragmentation.

from the ideal windshield's and/or wiper's shape due to production tolerances led to fast wear and a deterioration of the wiper's performance. The system "windshield wiper" has evolved into various adaptive systems. In one of the present systems, multiple pivots create a force-balancing action at several levels of the balancing arms. This balancing action results in the uniform distribution of pressure between the wiper blade and the glass along the wiper length, regardless of the glass shape.

Example 5.15
Another illustration of this line of evolution is the rapid development of keyboards for handheld computers over the last few years (see Chapter 2, Section *Separation of opposite properties in time*).

● *Increasing fragmentation*
Adaptability can also be enhanced by a transition from continuous, monolithic systems to systems consisting of several elements (fragments) (Fig. 5.9). For many systems, this involves the replacement of a monolithic object with a set of smaller objects: rods, pellets, fibers, powders, etc. The more disperse and fragmented the medium is, the more conforming it becomes.

Example 5.16
The design and fabrication of forming tools account for the majority of the time required for the production of a sheet metal part. These efforts can be justified only for mass production conditions. It is often very desirable to make small batches of formed sheet metal or plastic parts, however. A single universal stamping tool that can make parts of various shapes was conceived at MIT (Walczyk and Hardt, 1998) and developed in a collaboration between DARPA, Northrop Grumman Corp., MIT, and Cyril Bath Co. (Papazian, 2002). The tool (forming die; Fig. 5.10) has 2688 movable 21-inch-long pins with rounded tips. Eight pins make one module. The tool holds 336 modules that are installed and removed independently (Fig. 5.11).

Each module contains a motor, a microprocessor and motor controllers. The microprocessor receives instructions from the host computer, controls the motor and continuously stores information about the positions of each pin. The software divides the die into several zones and moves all of the modules in a zone at one time. Reconfiguration takes less than 12 min.

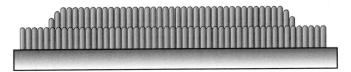

Fig. 5.10. Pins on this forming die produce a toroidal sheet with both negative and positive curves.

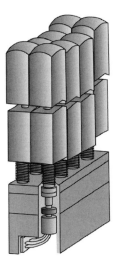

Fig. 5.11. Movable pins module.

5.4.2   Smart materials

The evolution toward increasing adaptability is greatly enabled by the growing use of so-called *smart materials* that involve devices that adapt to environmental requirements. Electrostrictive and magnetostrictive materials (see Chapter 3, Section *Modeling physical effects with sufields*), shape-memory alloys (SMA)[2] , electrorheological and magnetorheological fluids[3] are currently available smart materials (Banks *et al.*, 1996).

Smart materials may significantly enhance sensing, actuating and computation functions. When embedded into other systems, they provide for compact design, less complexity and higher reliability. Potential applications of smart materials are in monitoring and control systems for aerospace and marine vehicle structure, automotive monitoring and control devices, biomedical equipment and manufacturing process monitoring devices.

[2] Parts made of these alloys are able to change their shapes in response to temperature. These parts have a "memory" that allows them to be deformed into a temporary shape and then be restored to their original geometry by applying heat. SMA-based parts are often used in clamping mechanisms and actuators.

[3] These fluids can significantly change their viscosity when exposed to an electric or magnetic field. They can change from a thick fluid (similar to motor oil) to a nearly solid substance within a millisecond. The effect is completely reversed just as quickly when the field is removed. These fluids are being developed for use in automotive shock absorbers, engine mounts, surface polishing of machine parts, valves, clutches, etc.

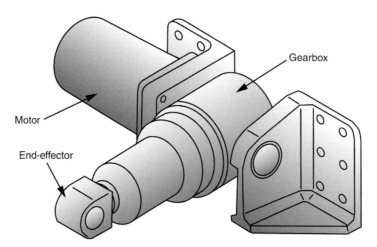

Fig. 5.12. Electromechanical actuator.

Example 5.17

One of these applications is the ongoing development of revolutionary *active aero-elastic wings* (AAW) by NASA, Boeing, and the US Air Force. These wings employ smart materials and actuators that help them reach maximum aerodynamic efficiency (Wlezien *et al.*, 1998; Cole, 2002).

The first plane built by the Wright brothers in 1903 had no ailerons or flaps to make turns and provide roll control. Since the plane's wings used fabric, the Wright brothers could control the wings' shape by "warping" the wingtips with cables. By pulling the cables, they twisted the wingtips either up or down, thus making the *Wright Flyer* turn.

As fabric in wings was replaced by wood and then by metal, ailerons, flaps, and leading edge slats were introduced (which corresponded to the "multi-joint" stage of increasing flexibility). These devices are usually controlled by hydraulic or electrome-chanical actuators, like the one shown in Fig. 5.12. Smart materials – specifically shaped memory alloys – allow the designers to return to the Wright brothers' idea – emula-tion of the bird's wings ("elastomeric" stage of increasing flexibility) – but on a more advanced level.

Two sets of shape memory wires (Fig. 5.13) are used to manipulate a flexible wing surface. Heating with an electric current controls the wires such that if the wing must be bent downwards, the wires on the bottom of the wing are shortened, while the top wires are stretched. The opposite occurs when the wing are bent upwards.

An aircraft equipped with AAW will be able to respond to constantly varying aero-dynamic conditions using its sensors to measure pressure over the entire surface of the wing. The response to these measurements will direct actuators that will func-tion like the muscles in a bird's wing, and alter the aircraft wing's shape to optimize its efficiency.

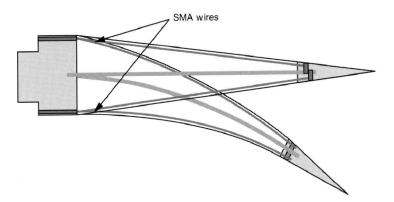

Fig. 5.13. Shape memory actuators.

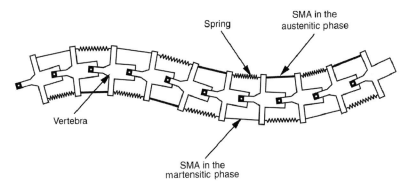

Fig. 5.14. Hydrofoil's "vertebrae column."

The savings in cost, weight and complexity (coupled to a much better performance) are obvious by comparing Fig. 5.12 and Fig. 5.13. Shape memory actuators (SMA) can also be used for hydrofoil control (Rediniotis, 2002). Figure 5.14 shows a part of the marine vehicle's "skeletal structure" similar to that of aquatic animals. Coordinated expansion and contraction (due to respective heating and cooling) of SMA "muscles" causes the rotation of the "vertebrae," which in turn causes a change in the shape of the hydrofoil.

5.4.3   Nonlinearity

One of the design principles that is useful for enhancing the flexibility of technological systems is a judicious use of *nonlinear effects* (Rivin, 1999). *Linear systems* are characterized by a proportionality of input and output quantities, i.e. between force F (input) and displacement Δ (output) in a mechanical spring, Fig. 5.15, as shown by line 1. This statement can be rephrased: the ratio of input and output quantities (stiffness $K = F/\Delta$ in case of spring in Fig. 5.15) is constant. The nonlinear systems are characterized by a more complicated dependency between input and output $\Delta = f(F)$.

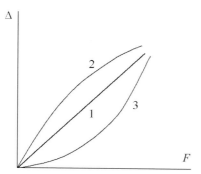

Fig. 5.15. Linear (1) and non-linear (2 and 3) spring constants.

Two simple embodiments of such dependency are shown by lines 2 and 3 in Fig. 5.15. In these cases, the ratio of input to output is not constant anymore. In the case of a nonlinear mechanical spring, the stiffness is $K = \mathrm{d}F/\mathrm{d}\Delta$ and is increasing with increasing F for the characteristic line 2 ("hardening" non-linearity), and decreasing with increasing F for the characteristic line 3 ("softening" non-linearity). Of course, combinations of linear, hardening, and softening segments in one load-deflection characteristic are also possible. If the shape of the nonlinear characteristics is appropriately designed, the system becomes adaptive since its output parameter would change in a desired manner with the changing input parameter.

5.5   Law of transition to higher-level systems

This law describes the evolution of technological systems toward increasing complexity and multi-functionality.

Law of transition to higher-level systems
   Technological systems evolve from mono-systems to bi- or poly-systems.

A *mono-system* is designed to perform one function. A *bi- (poly-) system* consists of two (more than two) mono-systems. These mono-systems can perform the same or different functions, or they can have the same or different physical properties. An example of a mono-system is a single lens. Two lenses constitute a bi-system (e.g., the simplest telescope, microscope, or eyeglasses), and three or more lenses make up a poly-system (a more advanced telescope or microscope).

One of the outcomes of this merging of mono-systems into a bi- or poly-system is the growth of system hierarchy, or development of new hierarchical levels; hence, the name of this law of evolution.

Transition to bi- and poly-systems is a major trend of evolution, and examples are in abundance (Fig. 5.16).

Fig. 5.16. Mono-systems evolve into bi- or poly-systems; bi-systems evolve into poly-systems, too.

- Bi-systems: scissors (grouping of two knives), spectacles, binoculars (pair of two telescopes), catamaran, two-barrel hunting rifle, semiconductor diode (combination of two semiconductor materials), candy with liquor; etc.
- Poly-systems: book, broom, multi-ink pen, roll of postal stamps, multi-drawer filing cabinet, transistor/thyristor (combination of three/four semiconductor materials), multi-cylinder internal combustion engine, multi-layer cake, etc.

The simplest way to form a more complex system is to combine two or more previously independent *mono-systems*. Bi- and poly-systems enhance the functional performance of each constitutive sub-system to such a degree that new and useful functional properties appear.

One of the advantages of the resulting complexity is the opportunity to distribute the functional load among the components. The load can be distributed *in space* and/or *in time*.

Example 5.18

An example of dividing the functional load in space is distributed computing. Probably the best-known example is the SETI@home project, in which thousands of individuals volunteer their computers to the Search for Extraterrestrial Intelligence (SETI) initiative. The goal is to split analysis of massive radio telescope data among numerous small CPUs (sub-systems). For comparison: the most powerful computer today, IBM's ASCI White, can perform 12 Teraflops and costs $110 million; SETI@home currently gets about 15 Teraflops and costs $500 thousand.

An example of distributing a load in time is a multi-stage rocket. Another example is the generation of high hydraulic pressures with a sequence of basic hydraulic pumps feeding one another so that each pump increases the pressure of the medium processed by the previous pump.

Bi-systems and poly-systems evolve along the same lines. The only difference between them is that a poly-system may create conditions for the development of an internal medium with new properties.

Example 5.19

The handling and shaping of thin glass plates is a delicate and time-consuming procedure. Gluing together a multitude of thin glass plates into a block solves this problem;

the block can be easily handled and machined without special precautions (U.S. Patent No. 3,567,547). In this example, combining the plates enables introduction of glue inside the system, to obtain not just a stack of plates, but a unified block ready for the process of shaping it (a polysystem). Applying glue to one plate would not achieve a similar effect. After the processing (e.g., machining to the required dimensions), the glue is dissolved and the separate glass plates are used as needed.

Technological systems may merge with each other to share their resources, increase functionality and the develop new systemic effects.

Example 5.20
A flexible manufacturing cell (FMC) consists of several similar or different machine tools (e.g., machining centers and/or turning centers) having a common material handling system and a supervisory computer. The effect of an FMC is much more than just putting together several machine tools. It can simultaneously machine several identical or different parts, or divide, in an optimized way, the machining and the measuring of one part between the component machine tools, etc.

Example 5.21
A current trend of PC evolution is the networking, or delegating of many functions to the servers and, essentially, returning (at a new level) to the use of individual PCs as "intelligent terminals." One can remember that in the pre-PC era, mainframe or mini-computers had all their computing power concentrated in the central CPU, and had several "intelligent terminals."

5.5.1   Lines of transition to higher-level systems (lines mono–bi–poly)

Increasing diversity of components

The taxonomy of bi- and poly-systems is shown in Fig. 5.17.

Homogeneous bi- and *poly-systems* have identical components (e.g., eyeglasses; hairbrushes). *Bi-* and *poly-systems with shifted properties* have similar components that are different in some aspects: size, color, shape, etc. (e.g., a comb with two sets of different density pins, a box of differently colored crayons). *Heterogeneous bi-* and *poly-systems* contain diverse components performing different functions (e.g., Swiss army knife, tool box), and *inverse bi-* and *poly-systems* contain components with opposing functions and/or properties (e.g., pencil with eraser; eyeglasses with sunshields) (Fig. 5.18).

The effectiveness of bi- and poly-systems is enhanced if their components are diverse. Diversity often creates new systemic effects. For example, bi-metal strips composed of metals with different coefficients of thermal expansion demonstrate large bending deformations for small changes of temperature. This effect is not present in single metal strips.

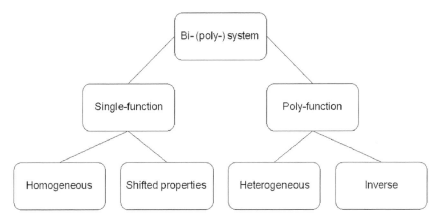

Fig. 5.17. Taxonomy of bi- and poly-systems.

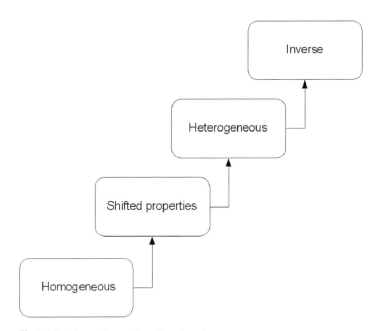

Fig. 5.18. Line of increasing diversity of components.

The minimal diversity is achieved in homogeneous bi-systems with shifted properties. For instance, the most primitive telescope (or microscope) is a bi-system of two
lenses with shifted properties. A more complex optical instrument is a poly-system of
lenses with shifted properties. Other examples are: a set of wrenches for hexagonal
nuts of different sizes, a set of wrenches for hexagonal nuts and Allen bolts, a universal screwdriver with tips of different shapes (straight, Phillips, etc.) and sizes in its
handle.

Example 5.22

A poly-system with shifted properties is used for suppressing helicopter noise (Edwards & Cox, 2002). As the commercial use of helicopters grows, so does the market sensitivity to helicopter noise. The foremost source of this noise is the main rotor carrying the blades. The recently proposed approach – *modulated blade spacing* – results in a dramatic reduction of rotor noise. Contrary to the conventional design, the blades are spaced unevenly. With even blade spacing, the frequency of the noise peaks (acoustic interaction between the blade and the helicopter body protrusions) is the same for each blade, and these peaks combine to make a high level of noise. With uneven or modulated blade spacing, the peaks are shifted and the acoustic energy is spread out over time.

Similar techniques are used for chatter/vibration suppression in milling cutters, saws, etc.

Heterogeneous bi- and *poly-systems* are composed of diverse components performing *different functions*, e.g., a mouse pad with a calendar, a coffee brewer with a clock and a radio, a wristwatch with a calculator, a key ring with a pen knife, etc. Many of these functions are *secondary*; they enhance the customer value of the system, or support its primary function.

Two conditions often lead to the assimilation of new diverse components (i.e., secondary functions).

First, after the viability of a new system has been established, its further deployment proceeds by the absorption of additional sub-systems. The goal is improved performance of the system's primary and secondary functions.

Example 5.23

The structure of early NC machine tools included a conventional machine tool, stepper motors for the feed drives, power amplifiers, and perforated tape readers. The process of feeding the tool to the preprogrammed coordinates was slow, since the positions of the points on the workpiece had to be precomputed and punched on the tape in a binary code; the accuracy was also low since no feedback system was used.

Major new mechanical subsystems were friction-reducing devices for feed drives and guideways, and feedback sensors and servo-control systems. They greatly enhanced the accuracy of machining. Some new computational devices were interpolators for automatic generation of the coordinates of intermediate points for the specified lines between the programmed nodal points. Later, the interpolators were replaced by computer-based controllers that reduced the time and effort needed for programming the sequence of feed motions necessary to machine a part. Other added subsystems were: a tool-changing module (for completing the machining sequence without moving the workpiece between several machining operations [e.g., milling, drilling, boring, tapping]), a part-loading robotic arm, and a subsystem for measuring the coordinates of selected points on the surface after machining.

Second, the resources for enhancing the primary function are exhausted or further enhancement of this function does not seem feasible and/or cost-effective.

Example 5.24

Watches have always been based on some periodic processes, however these basic processes have changed during the process of evolution: rotation of the Earth – sun dial, swinging of a mechanical pendulum – mechanical and electro-mechanical watches (e.g., "grandfather clock"), vibration of tuning forks or quartz plates – electronic watches, vibrations of electrons inside atoms – for scientific, "atomic" clocks. The performance of modern electronic wrist watches far exceeds the practical needs of watch wearers (errors in seconds per year). As a result, the development of wrist watches is heading toward incorporating additional – technical and non-technical – functions. One of these functions is combining timekeeping with an artistic role (diamond-studded watches, Mickey Mouse and Swatch watches, etc.). Other functions include: calendar, stop watch, the time in different time zones, notebook, alarm, calculator, radio, TV, emergency transmitter, blood pressure and/or blood sugar sensor, etc. – the number is growing incessantly, and will soon include GPS and cell phone subsystems.

The accelerated growth of both the number and the level of performance of additional functions is a reliable indicator that the system has reached its peak potential and is ready to be replaced by a new one.

The utmost manifestation of diversity is an *inverse bi-* or *poly-system*, which is a combination of components with opposite functions or properties. A classic example of an inverse bi-system is a pencil with an eraser. Another example is a semiconductor diode whose ability to rectify electric current is due to its composition of two semi-conductor materials with opposite electrical characteristics, one having an excess of electrons, and another with a surplus of positively charged "holes."

In some case, the principal advantage of inverse bi-systems is the capability to cancel harmful or undesirable properties of the components.

Example 5.25

Structural concrete blocks are economical; they can accommodate high compressive forces but are not suitable for tensile loading (i.e., bending). Steel rods are more expensive; they have very high tensile strength but are prone to buckling under compressive loading. They are also prone to corrosion. Concrete blocks pre-compressed by pre-tensioned embedded steel rods combine excellent strength for various loading patterns with high corrosion resistance (since the steel rods are protected from the environment by the concrete) and with relatively low cost and weight.

A unity of opposites may frequently result in elegant solutions to engineering challenges.

Example 5.26

At a metallurgical plant, slag slurry was pumped through a pipe. The slag particles gradually deposited on the pipe walls, thus reducing the pipe cross-section and obstructing the flow. Various solutions to alleviate the problem (using special coatings on the walls, the periodic removal of the deposit by mechanical or chemical means, replacement of the clogged pipe sections, conveying the slag in trucks, etc.) were expensive, ineffective, or both.

The problem was resolved by combining a system with its "anti-system." At the same plant, the coal waste, which is a by-product of coke production, was also conveyed through a pipe. In that case, however, coal chunks wore away the pipe walls. It was suggested to periodically switch the pipes used for conveyance of these materials. First, the slag flow would form a protective crust on the pipe walls, and then the coal chunks would remove this deposit.

• *Convolution*

Transition to bi- and poly-systems results in increased functionality, but this may come at the expense of increased complexity (two or more components instead of one). In other words, transition to bi- and poly-systems may not necessarily increase the degree of ideality.

Degree of ideality of bi- and poly-systems increases in the process of their *convolution*. The first step in this process involves the elimination of redundant auxiliary components. For example, a double-barreled gun has only one butt. Such systems are called *partially convoluted bi-* and *poly-systems.*

Example 5.27

Machining complex parts on computerized numerical control (CNC) machining centers sometimes involves the first machining operation, then the transfer of the part to a CNC coordinate measuring machine (CMM) that allows one to identify dimensional deviations caused by cutting forces, tool wear, thermal deformations, etc. After the measuring operation, the part is transferred back to the machine tool for the corrective machining. Combining the functions of the machining center and CMM by using a measuring probe on the machine tool results in significant savings since no back-and-forth part transfer is needed. Of course, some modifications of the machine tool allowing for accurate repeatable positioning of the probe in the spindle may be required.

Example 5.28

Portable products such as cell phones, personal digital assistants (PDA's), and the like have become increasingly multi-functional. The need for increased functionality demands more memory chips on a single printed circuit board (PCB). This, however, compromises ever tightening packaging constraints.

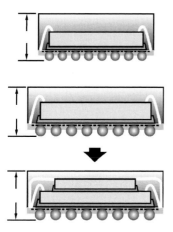

Fig. 5.19. Two-storey IC chip package.

A partially convoluted system of a stack chip packaging should resolve this conflict (at least for a while [Kada & Smith, 2000]). The idea is to save room on a PCB by stacking up chips vertically and interconnecting them inside a single very compact package (Fig. 5.19). A "two-storey" chip stack saves up to 57% of the PCB area.

Further evolution usually leads to the development of *completely convoluted bi-* and *poly-systems*. In these systems, one component may perform two or more functions.

Example 5.29

An example of a completely convoluted bi-system is the protective suit for emergency rescue workers in mines developed by Genrikh Altshuller (Altshuller, 1988). A protective suit must have a cooling system and an oxygen-supply system. Altshuller developed a suit that combines these two systems. The key to the solution is using liquid oxygen. Firstly, oxygen is used in the cooling system, and once its low temperature has increased during the cooling process, it is used for breathing. This combination has resulted in significant weight reduction and allowed for a substantial increase of the system resources between recharges of the life support gear.

Another example of successive convolution is the evolution of the kitchen garlic press.

Example 5.30

Two separate mono-systems
Garlic cloves are reduced to pulp for cooking by squeezing the clove in the press (mono-system A) through a perforated plate (Fig. 5.20). Cleaning the plate is a tedious and unpleasant procedure usually performed by piercing the holes in the plate, clogged with garlic residue, with a sharp object, e.g., a pin (mono-system B).

The first stage of transition to a poly-system is the combination of a number of pins into a cleaning punch to clean all the holes in the plate simultaneously (Fig. 5.21). The

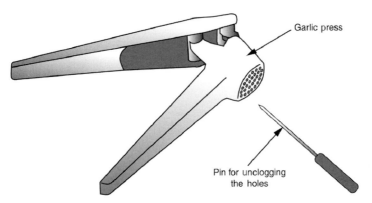

Fig. 5.20. Garlic press and a pin for unclogging the holes.

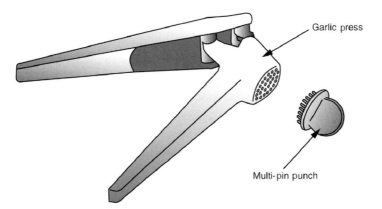

Fig. 5.21. Multi-pin system.

created poly-system is partially convoluted, since one punch controls a multitude of pins. This poly-system can be considered as a new mono-system. Usually, a multi-pin punch is sold together with the garlic press in a set.

Partially convoluted bi-system
In the next step of evolution, the multi-pin punch is attached to the press, Fig. 5.22.

Completely convoluted bi-system
Finally, the multi-pin punch is integrated with the press, Fig. 5.23. The perforated plate is attached to one lever of the garlic press, while on the back side of the other lever is a set of pins which correspond to the holes in the plate when the levers are reversely folded. The pins clean all the holes in one stroke, while the production costs are reduced, because the lever with the cleaning pins can be molded as one part. The result of the evolution process is one device (a higher-level system) of about the same size and price as a conventional garlic press, but which replaces two devices and performs the cleaning function much faster and much more conveniently.

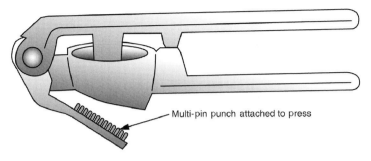

Fig. 5.22. Partially convoluted pins/punch.

Fig. 5.23. Completely convoluted pins/punch.

Completely convoluted bi- and poly-systems become, in their turn, mono-systems and may again evolve by merging with other mono-systems. This evolutionary spiral is shown in Fig. 5.24.

5.5.2   Evolution of materials

Many technological systems originate as homogeneous materials (substances). The interaction of a homogeneous material with other materials (systems) usually reveals that, in order to function better, it should possess different, occasionally contradictory, physical properties; it may be required to be – in various areas – hard and soft, solid and gaseous, etc.

Increasing non-uniformity of materials

Different demands to different areas of the material lead to its segmentation. Each segment – a *functional zone* – may have its own physical properties and thus can perform

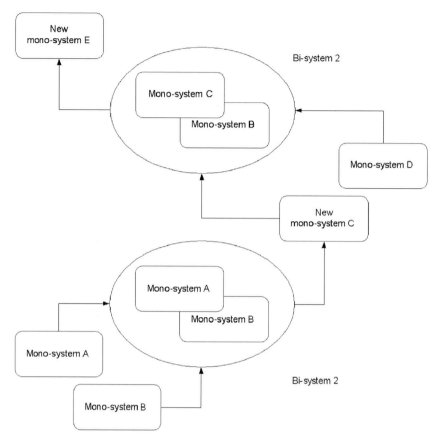

Fig. 5.24. Convolution spiral.

a different function. Separation into functional zones entails specialization – each zone (i.e., substance structure) performs one function. Development of functional zones may begin within the boundaries of a *mono-material* part. New properties (functions) are attained by drawing on internal resources of the part: different thermal and/or mechanical treatments (cold or hot working) of a certain area, changing the shape of a certain section of the part, its density (e.g., by introducing holes or pores), color; etc. If using internal resources of the material cannot produce the necessary effect, then functional zones are formed by adding other materials with the sought after properties, thus creating *bi-* and *poly-material* structures (i.e., composite materials). Thus, a solid part may become a bi- or poly-system. As should be expected, composite materials may eventually convolve into another mono-material that has built-in functional zones due to its molecular or atomic structure.

Example 5.31

The emergence of functional zones by shaping is clearly seen in the early evolution of the hammer (stages 1–5; Fig. 5.25 [Bassala, 1988]): first – a piece of unprocessed stone, then – the gradual formation of a head, after that – the shaping of a handle, etc.

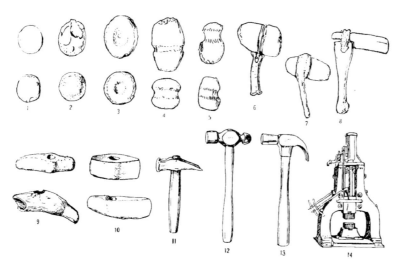

Fig. 5.25. Evolution of the hammer.

Example 5.32
Another example of the evolution into bi- and poly-systems of a solid part is the development of automotive headlamp lenses; from the flat and smooth protective glasses of the Model T, to the curved prismatic lenses of today's autos performing several functions (focusing, deflecting, condensing light, and protecting the bulb and reflector).

Segmentation into functional zones often starts with differential treatments of a uniform material in different areas.

Example 5.33
A typical example is the surface heat treatment (such as induction hardening) of steel products. In power transmission gears, there is a need to have hard surfaces of teeth that are subjected to high contact loads while sliding against counterpart teeth. On the other hand, the core of the tooth should have a relatively low hardness, since high hardness is associated with brittleness under shock loads in the transmission.

Example 5.34
The use of different materials in different functional zones can be illustrated by the use of coatings. A house is built of wood, which has good structural properties but deteriorates under the influence of environmental elements. Coating the wooden structure with another material – protective paint or vinyl boards – makes the system "house" perform much better over time.

Coatings are extremely popular since they allow for the utilization of unique properties of advanced expensive materials without incurring prohibitive costs. Frequently, these advanced materials cannot be used in bulk because of their poor structural properties. Examples include cutting tools coated with extremely hard materials (such as titanium carbide, titanium nitride, diamond, etc.), structural components coated with low-friction materials (reducing their friction coefficient from 0.15–0.2 down to 0.03–0.05), structural parts made of strong but not temperature-resistant materials (coated by ceramics), etc.

Occasionally the line of increasing non-uniformity branches into several lines: An added material – essentially a subsystem – is deployed into a mono-system. This mono-system itself is destined to evolve along the line "mono-bi-poly."

Example 5.35
The game of golf is believed to have originated in Scotland in the sixteenth century (McGrath & McCormick, 2002). The first golf clubs and balls were made from wood. The predecessor of modern golf balls, the Gutta Percha ball, or 'Guttie,' was invented in 1848. It was made from the rubber-like sap of the Gutta tree which grows in the tropics. While the Gutta Percha ball gave players more opportunities to control the game, the multi-piece ball introduced in 1898 by Coburn Haskell gave the average golfer an extra 20 yards from the tee. This ball had a solid rubber core wrapped in rubber thread sheathed in a Gutta Percha shell.

Rubber – a subsystem of the Haskell ball – convolved into a mono-system with the introduction of the so-called one-piece ball in 1966 (Fig. 5.26). The one-piece synthetic rubber ball is very durable and economical but it does not offer a long distance and "soft-feel" (greater control). To overcome these deficiencies, Bridgestone brought in a two-piece ball in 1985. That ball has a solid synthetic rubber core and a surlyn cover for durability. Later, Bridgestone further evolved the golf ball along the path of increasing non-uniformity and introduced three- and four-layer balls in which the core material has multiple wrappings.

Development of integral functional zones (convolution) can be achieved by changing the physical and/or chemical properties of the material in those particular areas.

Example 5.36
The differential hardening of the gear teeth can be realized by using low carbon content (non-heat-treatable) steel that has high ductility, and by modifying it by increasing the carbon content in the external layers (carburization process). After carburization, the whole gear is heated and quenched, but only the carburized external surfaces of the teeth are hardened during this process. The effect is similar to the above-described induction hardening, but less expensive alloys can be used for gear manufacturing and less expensive processing equipment is required.

Further opportunities for complete convolution of materials will most likely be linked to advancements in *nanotechnology* (Drexler, 1992; Merkle, 1997). This technology allows a product to be assembled from individual molecules or atoms, or microstructures

Fig. 5.26. Evolution of the golf ball.

like carbon nanotubes. It was first proposed by Richard Feynman (Feynman, 1959), and is now being actively pursued by many commercial and academic organizations. In nanotechnology, both a combination of atoms according to the needed properties of the product, and their geometric pattern (shape) are controlled, and can be individually tailored to manufacture custom-made products in one copy or in quantities.

These potential developments bring us to another law of evolution.

5.6   Law of transition to micro-levels

All technological systems are composed of substances. Any substance is a multi-level hierarchy of various physical structures. It is sufficient, for practical purposes, to present this hierarchy as consisting of the following levels/physical structures (Fig. 5.28):
• Crystal lattices, molecular aggregates
• Molecules
• Atoms, ions
• Elementary particles.
Any physical structure inhabiting some level in the hierarchy is a *micro-level structure* for all the structures that occupy higher levels (*macro-levels*). In the process of evolution,

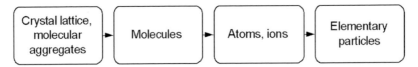

Fig. 5.27. Line of transition to micro-level.

technological systems make transition from macro- to micro-level. This means that functions ordinarily performed by such "macro-objects" as shafts, levers, gears, and the like become gradually assigned to "micro-objects:" molecules, atoms, ions, and elementary particles.

The *transition to micro-level* is a process of employing lower-level physical structures. Various conflicts that develop during the process of evolution can be resolved by the transition into micro-level.

Law of transition to micro-level
  Technological systems evolve toward an increasing use of micro-level structures.

5.6.1 Line of transition to micro-levels

Example 5.37
This line can be illustrated by the evolution of metal machining technologies (Fig. 5.28).

Crystal lattice
Traditional machining technologies (milling, drilling, grinding, etc.) operate at this level. In these technologies a cutting tool (which is harder than the removed material) is forced against the material to be removed.

Molecules
In chemical and electrochemical machining, metal is removed by the controlled etching reaction of chemical solutions on the metal.

Atoms, ions
Atomic-level tools are used in the process of plasma arc cutting, in which a high-temperature plasma cuts the workpiece by melting it. The tool in electro-discharge machining operates at the same level, wherein the workpiece is eroded by high-energy electrical discharges.

Elementary particles
These are employed in laser and electron beam machining (vaporizing the material with an intense beam of light from a laser and with a beam of high-velocity electrons, respectively).

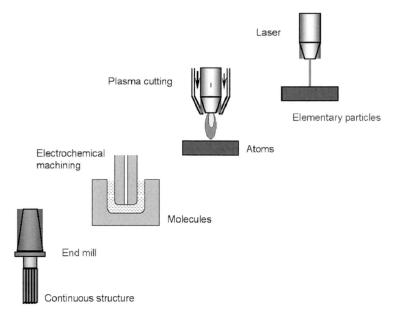

Fig. 5.28. Evolution of metal machining technologies.

Transition to micro-level involves two principal mechanisms.

(1) Micro-level structures take over functions that used to be performed by macro-level structures (e.g., replacement of conventional mechanical cutting tools with photons emitted from a laser).

(2) Micro-level structures control the physical properties and behavior of macro-level structures. Eyeglasses with photochromic lenses employ such a mechanism. These eyeglasses eliminate the need for sunglasses or sunshields because they darken when exposed to bright light. This change in translucency is due to crystals of silver chloride that are added into the glass during manufacturing. Light interaction with a chloride ion produces a chlorine atom and an electron. This electron then unites with a silver ion, thus producing a silver atom. Numerous silver atoms cluster together and block the incident light, causing the whole lens to darken. The degree of "darkening" is proportional to the intensity of the light.

5.7   Law of completeness

This law establishes the number and functionality of the principal parts of any autonomous technological system.

Law of completeness
> An autonomous technological system consists of four principal parts:
> working means, transmission, engine, and control means.

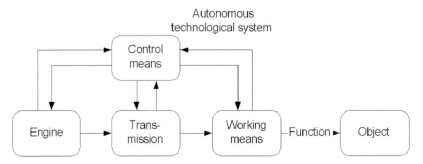

Fig. 5.29. Structure of an autonomous technological system.

Fig. 5.30. Successive dislodging of "human components" from technological systems.

A working means is a component that directly performs the primary function of the system. A transmission transforms the energy produced by the engine (energy source) into the energy that can control the working means. A control means allows for changing the parameters of the other principal parts (Fig. 5.29).

Example 5.38
Consider, for example, the technological system "photographic camera." The purpose (primary function) of the camera is to transmit the light reflected from an outside body to the film (or to the sensor plate in a digital camera). Therefore, the object to be manipulated by the camera is the light, the working means is the set of lenses, the focus-adjusting mechanism is the transmission, the electric motor is the engine, and the control in many modern cameras is performed by a microprocessor.

The milestones in the evolution of the hammer – from a stone to the sledged hammer – in Fig. 5.25 clearly show the gradual emergence and development of the principal parts of this technological system.

5.7.1   Line of completion (dislodging of human involvement)

At early stages of development of many technological systems, some of their principal components are substituted by "human components". As these systems evolve, the "human components" are gradually eliminated, and the functions they perform are delegated to technological systems. This elimination of human involvement follows the line shown in Fig. 5.30 (Altshuller *et al.*, 1988).

Example 5.39
The first windshield wiper was invented in 1903. In this wiper design the rubber blade was rotated manually across the windshield. Manual wipers (operated either by the driver or by a passenger) were used on military Willys Jeeps as late as WWII (McNeil, 2002). The next wiper design was powered by a vacuum pump linked to the car's engine. The disadvantage of that design was the dependency of its speed of operation on the speed of the vehicle. This eventually led to the replacement of the vacuum pump by a motor. The next group of major innovations involved changes to the wiper control system. These included intermittent wiper design, rain-sensing wipers, etc.

Example 5.40
The precursor of a modern toothbrush is a toothpick made from a wooden thorn or a bird's feather. This toothpick performed the functions of both the working means and transmission, while the user's hand served as the engine and control means. Egyptians used such tooth cleaning implements some 5000 years ago. They, and other ancient peoples, also used twigs that were smashed at one end to increase their cleaning surface.

Tooth cleaning became much more effective when the working means, i.e. the toothpick's tip, evolved into a poly-system, i.e. a set of bristles. This development triggered the deployment of the rest of the toothpick into a more user-friendly "transmission", i.e., the handle. The first bristled toothbrush is believed to have been invented in China in the fourteenth century and was brought to Europe three centuries later. It used boar hairs for bristles and a bone for a handle.

During the next few centuries, the toothbrush's history was quite uneventful (the only noticeable changes were the introduction of celluloid handles in the 1920s, and the replacement of boar-hair bristles by nylon ones in the 1940s).

The major breakthrough came with the deployment of the toothbrush "engine" in 1956. That year, the first successful electric toothbrush – Broxodent – was introduced in Switzerland by Squibb (the first electric toothbrush was marketed in the US in 1880, but it could not conquer the market [Travers, 1995]). In 1961, General Electric introduced a rechargeable cordless model.

5.8   Law of shortening of energy flow path

One of the necessary conditions for effective functioning and controllability of energy-transforming technological systems is passage of energy through the system to its output (working organ). Applying this statement to the development of actual systems, it is useful to distinguish between two basic classes of design problems. One class is the problems that go along with *changing* a system (synthesis of a new one, improvement

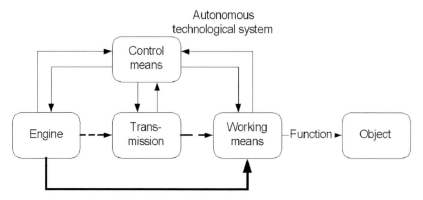

Fig. 5.31. In the most effective systems, the source energy is applied directly to the working means.

of an existing one). The second class is *measurement problems*, in which the goal is to detect, or to measure, or to monitor certain parameters of the system.

   In problems of the first class, the energy passes from the source (e.g., combustion chamber in a gas turbine installation, boiler in power generating stations, driving motor in processing machines, etc.) via intermediate sub-systems (steam piping and steam turbines in power generating stations, mechanical transmission in processing machines) to the working means (electric generator in power generating stations, spindle with cutting tool in mining and metal cutting machines), and to output systems (electric power distribution network for power generating stations, cutting process in mining and metal cutting machines).

   In problems of the second class, energy emanating from the object being measured/monitored should be sensed so as to obtain the required information. In such cases, energy passes from the object to the sensor (transducer), then it is transmitted via certain auxiliary lines (electrical conduits for transducers with electric current as output, piping for pneumatic transducers, etc.) to a signal conditioner from which it is further transmitted to an indicator, a data acquisition system, or a feedback unit (for servo-controlled systems), to be transformed into information.

Law of shortening of energy flow path
   Technological systems evolve in the direction of shortening of energy flow passage through the system.

Usually, this evolutionary process is associated with at least one of the two following tendencies (lines of evolution) (Fig. 5.31).
(1) Reduction in the number of energy transformation stages (both the transformation between different forms of energy, such as electrical and mechanical, mechanical and thermal, etc., and the transformation of energy parameters, such as mechanical

transmissions transforming speeds and torques/forces between input and output members).

(2) Proliferation of such forms of energy which are easier to control. These forms can be listed in the order of increasing controllability:

- gravitational
- mechanical
- thermal
- electromagnetic.

This law of evolution can easily be recognized by tracing the evolution of many major technological systems.

Example 5.41

The majority of modern passenger trains use diesel locomotives. In these locomotives, thermal energy is transformed into mechanical energy in the diesel engine, then parameters of the mechanical energy (speed and torque) are transformed again by a mechanical transmission to drive the wheels with the required speed. The wheels transform rotational mechanical energy into the train's translational energy. The engine's mechanical transmission is needed in order to run the diesel engine at its optimal conditions of speed, torque, and efficiency. Such locomotives are being gradually replaced on the most intensely used routes by electric locomotives in which the source of energy is electric current, whose parameters in some cases are transformed on-board by highly efficient transistor-based systems. Then, electric motors directly, or via simple mechanical transmissions, drive the wheels. This process improves both energy efficiency and performance.

The next stage is the Maglev (magnetic levitation) system, in which the propulsion is provided by linear electric motors, thus ultimately shortening the path of energy transmission. The wheels in the conventional train not only provide propulsion, but also support the train on the ground. A magnetic levitation bearing system reduces friction and energy losses, reduces noise and provides suspension action (vibration protection).

Example 5.42

At the beginning of the twentieth century, machine shops had centralized drives in which an internal combustion engine or an electric motor drove long transmission shaft(s) which hung over the whole shop (Fig. 5.32). Numerous belts transmitted mechanical power from the transmission shaft to individual machine tools. Within each machine tool, this driving input was transformed by other mechanical transmissions into the required parameters of energy (spindle and feed motions). Other internal transmissions provided changing speed ratios between the spindle and the other units (feed motion of the tool support carriage, indexing motions in gear cutting machines, etc.). In the

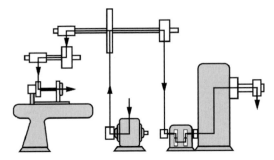

- Central motor/internal combustion engine
- Overhead transmission shaft
- Belt drives to spindles of individual machines
- Internal transmissions to feed motions

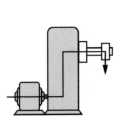

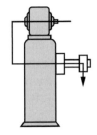

- Individual electric motor
- Mechanical transmissions to spindle
  and feed motions

- Separate motors for motions not
  precisely tied to spindle motion

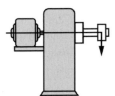

- Separate electric motions synchronized
  by electric means

- Separate electric motors integrated by
  "electronic gear box" (not shown)

- Spindle serves as shaft of spindle motor
  (not shown)

Fig. 5.32. Evolution of machine tool transmissions.

1930s, the energy transmission chain within the machine was shortened by installing individual electric motors on each machine tool.

All the internal transmissions remained, however. Some machine tools for machining complex parts (e.g., for spiral bevel gears) had half a dozen internal transmissions – each of which had to be precisely fabricated and tuned for each set of gear parameters in order to accurately machine complex surfaces. Starting in the late 1930s, many systems (such as grinding machines' feed systems) were detached from the main power input, and the energy transmission paths were further shortened. Initially, these independent systems did not involve precision motions, which were still synchronized by mechanical transmissions from the single motor. In the late 1950s, however, a high

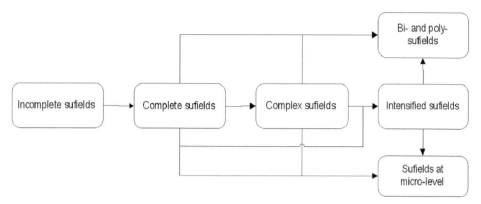

Fig. 5.33. Lines of evolution of sufields.

precision gear grinder was developed by Reishauer Co. in Switzerland in which all motions were performed by individual motors directly connected to output mechanical elements (motion of the workpiece spindle and motion of the grinding wheel spindle), with the motors synchronized by an electronic system.

This direction further developed into an "electronic gear box" which can provide accuracy of motion synchronization within fractions of an angular second. Such accuracy is very difficult, if at all possible, to achieve by mechanical means. Modern CNC gear cutting machines have all the numerous motions necessary to generate accurate spiral bevel gears performed by individual electric motors directly, without mechanical transmissions. General purpose machine tools also have separate direct drives (motor → ball screw → moving unit, without gear boxes) for all axes of motion. The increasing requirements for velocities, accuracy, and controllability of translational motions led to a wide use of linear electric motors, thus completely eliminating mechanical transmissions and shortening the energy path to the ultimate. The most stubborn system was the spindle drive which required a very broad range of speed variation (ratio of max. to min. speed 100–1000) while having high installed power, up to 30–70 kW for medium-sized machine tools. This lead to the use of mechanical transmissions (belt drive with gear box) for the spindle drives. The modern high speed/high power machine tools, however, have a direct connection between the spindle and the electronically controlled variable speed motor. More and more often, the spindle is totally combined with the motor (the spindle is the rotor).

5.9   Law of increasing controllability

This law is also called the *law of increasing substance–field interactions.* It delineates the evolution of sufields (Fig. 5.33).

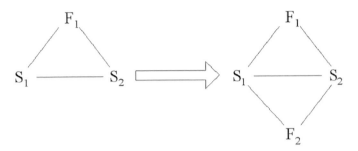

Fig. 5.34. Double sufields.

Fig. 5.35. Chain sufields.

$$C = \frac{\text{Number of sufields}}{\text{Number of sufield elements}}$$

Fig. 5.36. Coefficient of convolution.

Law of increasing controllability
Technological systems evolve towards enhancing their substance–field interactions; this involves the following.
• Incomplete sufields develop into complete sufields, which then evolve into complex sufields.
• Complete and complex sufields develop into intensified sufields.

5.9.1   Complex sufields

Often, there is a need to maintain an existing sufield (function) and, at the same time, to introduce a new, typically auxiliary function. This is usually achieved by a transition to *complex – double* (Fig. 5.34) or *chain* (Fig. 5.35) – *sufields.*

Transition to complex sufields is another mechanism of the increasing degree of ideality. An increase in the number of elements in complex sufields is paid off by increased functionality of these sufields vis-à-vis simple ones. The coefficient of convolution, *C*, (Vertkin, 1984), is the ratio of the number of sufields describing the system, to the number of elements in these sufields (Fig. 5.36).

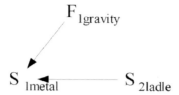

Fig. 5.37. Poorly controlled sufield for discharging molten metal.

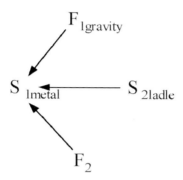

Fig. 5.38. Adding a control field.

In fact, this coefficient allows one to measure, to a certain degree, the change in the degree of ideality. For a simple sufield, $C = 1/3$, for a double sufield, $C = 1/2\,(2/4)$ and for a chain sufield, $C = 2/5$.

Example 5.43
Molten metal is poured into the mold through an opening in the bottom of the ladle. The rate of discharge is proportional to the level of the metal in the ladle: the lower the level, the lesser the hydrostatic pressure is, and the slower the flow of metal. Controlling the rate of discharge by varying the diameter of the opening is a difficult task.

Sufield analysis
The IMS (Fig. 5.37) is a complete sufield. There are no well-controlled elements in the sufield, however. Therefore, the whole sufield is not well-controlled. To enhance the sufield's controllability, a controlling field, F_2 (Fig. 5.38) is added into the system.

Solution
The metal is spun in the ladle with a rotating magnetic field; this allows control of the level of the metal in the ladle.

Example 5.44
How to increase the effectiveness of a hammer?

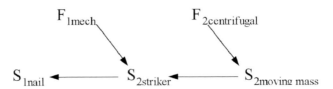

Fig. 5.39. Sufield model of a dynamic hammer.

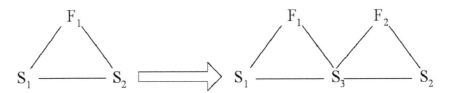

Fig. 5.40. Evolving the link S_1–S_2 in chain sufields.

Sufield analysis

The effectiveness of a hammer increases with the mass of its striker. Use of a heavy hammer is rather tiresome, however. Thus, there should be an additional mass in the striker, and it should be absent. This physical contradiction can be resolved by separating the opposing demands in time: the additional mass is absent before the strike, but it appears in the striker at the moment of the strike. This prompts an idea of making this mass moving and controlling it with centrifugal forces (Fig. 5.39).

Solution

A hollow striker is connected to a hollow handle that is filled with a heavy granular medium (e.g., lead shots). The shots are conveyed inside the striker cavity by centrifugal forces just before the strike, thus resolving the conflict. An additional advantage of such a design is reduced noise level due to the vibration damping action of the granular medium.

A chain sufield can also be developed by deployment of links within an existing sufield. In this case, connection $F_2 \rightarrow S_3$ is built into the link $S_1 - S_2$ (Fig. 5.40).

Example 5.45

US Patent 6,032,772 describes a viscous fluid clutch for controlling the rotation of an engine cooling fan in automobiles. The clutch includes a rotor attached to the engine output, a stator attached to a fan assembly and an electromagnetic coil attached to the stator. Both the gap between the rotor and the stator and the gap between the rotor and the coil are filled with a magnetorheological fluid. The coil creates a magnetic field in the gaps to vary the viscosity of the magnetorheological fluid, thus providing a variable driving force dependent on the strength of the magnetic field.

5.9.2   Intensified sufields

Intensification of interactions among sufield elements can be achieved by the following means.

- Replacement of an uncontrollable or poorly controllable field with a better controllable field, e.g., replacement of a gravitational field with a mechanical field, of a mechanical field with an electric field, etc.
- Increasing the degree of flexibility of the sufield elements.
- Tuning or detuning the frequency of a field action with the natural frequencies of the object or the tool.
- Some others (*see* Standards of Group 2.2. in Appendix 3).

5.9.3   Bi- and poly-sufields; transition to a micro-level

At any stage of their evolution, all elements of sufields may evolve into bi- and poly-elements, as well as undergo transition to a micro-level. For example, if a toothpick is a mono-system, then a set of bristles in a conventional toothbrush is a poly-system, while an ionic toothbrush is a system operating on a micro-level.

5.10   Law of harmonization of rhythms

This law of evolution determines the most effective temporal structure of technological systems.

Law of harmonization of rhythms
> The necessary condition for optimal performance of a technological system is coordination of the periodicity of actions of its parts.

The most viable systems are characterized by such coordination of their principal parts when the actions of these parts support each other and are synchronized (harmonized) with each other. This is very similar to production systems in which "just-in-time" interactions between the principal parts of the production sequence result in the most productive operation. If the synchronization principle is violated, then the constitutive parts of the technological system may interfere with each other, and performance will suffer.

Application of this law to technology and product development can employ several approaches:

- Coordination of periodic motions and other parameters between the constitutive parts.
- Use of resonance.
- Abatement of undesirable periodic motions.

Coordination between parts of a system may not be associated with natural frequencies or other dynamic characteristics, but with the sequence of action of various parts. Properly implemented coordination can greatly improve the performance characteristics of the system.

Example 5.46

Conventional spray guns have valves controlling the flows of paint and air to the spray nozzle. They are synchronized in such a way that when the gun is triggered, the air valve is opened first, and when the painting process stops, the paint valve is closed first. However, this is accompanied by the "spitting" of dried paint drops from the nozzle and onto the painted surface. US Patent No. 4,759,502 suggests an improved synchronization routine, where the paint valve opens and closes shortly before the air valve, with the time delay being adjustable. This improved synchronization leads to a significant improvement of the painting process.

Use of vibrations to perform various production and measurement operations is widespread (so called "vibratory technology").

The effects of resonance and of standing waves are extensively used in various applications. They allow for sustaining vibrations with minimum energy consumption, and also assist with various production processes.

Example 5.47

The effect of a standing wave is used in a highly productive method to manufacture microspheres. Molten material is ejected at a high speed through a hole in a vibrating die. The frequency of vibrations is set as a function of the desired diameter of the sphere. Vibrations generate a standing wave in the jet stream, and the latter breaks into precisely identical segments, which immediately attain a spherical shape under the influence of surface tension forces. This method can produce up to a million spheres per second, with diameters ranging from 0.0004 in. (10 μm) to 0.04 in. (1 mm).

Since production processes are frequently characterized by continuously varying parameters, natural frequencies of the production system also fluctuate, thus deviating from the resonance with vibration excitation from the driving units. In other cases, the resonance can be an undesirable regime. Servo-controlled self-tuning (or self-detuning in the latter cases) systems are very useful in such circumstances.

Example 5.48

In ultrasonic machining (Fig. 5.41) (Astashev and Babitsky, 1984) a tool is attached to a vibrating resonator excited with ultrasonic frequency (20–100 kHz) by a magnetostrictive vibrator. The tool interacts with the workpiece inside the tank filled with an abrasive suspension. The tool unit is pressed to the workpiece with a certain feed

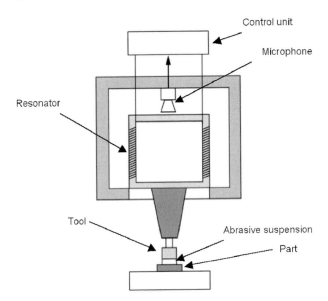

Fig. 5.41. Ultrasonic machining tool.

force. The tool vibrations are transmitted to the abrasive particles which impact the blank being machined. Even the hardest materials can be machined using this technology, and the most intricate shapes of the tool are mirror-formed on the part. The most energy-efficient machining process is realized when the system is at resonance with the forcing frequency of the vibrator. The system is precisely tuned before the machining (idle condition, when there is no contact between the tool and the machined blank). However, the natural frequency depends on interaction in the working zone and on the magnitude of the feed force. Due to the process related to detuning, the system always deviates from the resonance, resulting in a very low efficiency, about 3%. To solve this problem, a microphone is used as a feedback sensor. A control unit is continuously changing the generator frequency using vibration intensity sensed by the microphone as the feedback signal. This allows the system to continuously maintain the resonance conditions, thus greatly improving productivity and raising efficiency as much as 50%.

Example 5.49
A very different problem of unwanted vibrations during end milling on high-speed machine tools is solved using the same approach. In this case, a microphone picks up the vibration signal from the cutting zone, the control computer determines the frequencies of structural resonances and of self-excited vibrations, and if vibration amplitudes exceed the tolerance, the computer commands the CNC controller to change the spindle rpm. If several changes do not bring the process into compliance, the machine tool is stopped. Such a system greatly lengthens the life of the tool and improves the surface finish of the machined part.

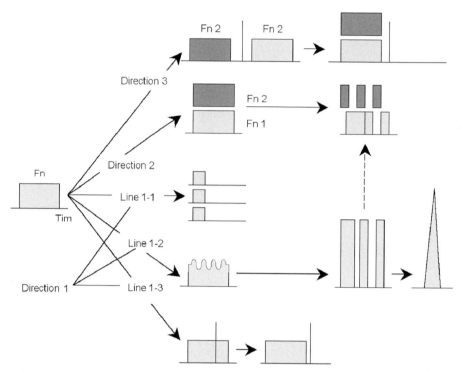

Fig. 5.42. Lines of chronokinematics.

This last example demonstrates the abatement of undesirable periodic motions. There are numerous techniques for vibration control, which are described in detail in the specialized literature. To eliminate resonance, either the forcing frequency of the exciter or the natural frequency of the vibrating system should be changed. In passive systems (not equipped with automatic controllers), the resonance or conditions for generation of self-exciting vibrations may develop at different regimes, at another natural frequency, etc. This can be avoided, in some cases, by eliminating periodicity, which can be achieved by making the cutting teeth on a milling cutter, or on a saw blade non-uniformly placed, or making the spindle RPM vary, thus breaking the periodicity of excitation. Another approach is to enhance the damping (energy dissipation) in the system, thus reducing the intensity of resonance.

5.10.1 Lines of chronokinematics

In order to increase its ideality, a system has to meet two conflicting demands: to become ideal, the system has to spend no time performing a function, but in order to do real work, some time has to be spent.

There are three directions of temporal idealization of systems (Fig. 5.42, where **Fn** stands for function). These directions are manifested in several lines of *chronokine-matics* (Fey, 1988).

- Direction 1: The system's function does not change, but the time needed to perform the function decreases.
- Direction 2: A system simultaneously performs several functions.
- Direction 3: A process of increasing the number of functions is accompanied by a shortening of the time needed to perform these functions.

Direction 1 combines three lines of chronokinematics.

- *Line 1: Distribution of a function among two or more simultaneously operating systems (or sub-systems).*

There is an abundance of examples in TRIZ literature illustrating this line (see, for instance, *Example 5.18*).

- *Line 2: Transition to a peak mode of function performance. This line is characterized by the succession of the following regimes: oscillations, pulsations, and peaks.*

Example 5.50

Radio engineering began with analog, i.e., continuous signals. A major breakthrough was the transition to using pulses. This enabled, in particular, inserting the pulses of one message in the pauses between the pulses of another message thus dramatically increasing the density/effectiveness of radio communications. The next big step was the development of signal compression systems used, specifically, by the military for the transmission of classified information. In these systems, a message does not proceed from a transmitter directly to an antenna, but is first accumulated in the memory device. After the memory is filled, the message is compressed and sent out in the form of a very short pulse.

Example 5.51

A conventional syringe for administering medical drugs has a barrel, a plunger and a needle. Drugs are supplied to the body part by moving the plunger. In many modern drug delivery devices, increasing the drug absorption by the body tissues is achieved by applying ultrasonic energy to the medicine (see, for example, Japanese Patent No. 2152466A2). In more advanced devices, the needle is absent. Needleless transdermal syringes deliver drugs under high – 40–65 bar – pressure; a narrow jet of drug runs through the skin at a speed of up to 500 m/s (see, for example, US Patent No. 6,168,587).

- *Line 3: Transition to preliminary action. This line features two major regimes of functioning: partial preliminary action and complete preliminary action.*

Example 5.52

At the dawn of software engineering, a programmer had to write a new program for each particular occasion. This was very time consuming and, naturally, led to the development of routines – codes for certain repeating or typical applications (e.g., trigonometric functions) written in advance. The next logical step was the development of CASE

(Computer-Aided Software Engineering) systems that allowed for the practically complete automation of writing complex codes for specific application domains (e.g., for generating various financial reports). The user of such a system needs only to describe the application (report) to be developed, and the system will automatically generate a code that will produce this application.

Direction 2 is well illustrated by examples on transition to convoluted bi- and poly-systems.

Direction 3 integrates two approaches to increasing ideality: transition to preliminary actions and transition to bi- and poly-systems.

Example 5.53

Any part utilizing a shape memory alloy can alternate between two configurations/functions. These configurations/functions are pre-programmed into the part's substance structure.

5.11   Interplay of the laws and lines of evolution

One could develop a perception that the laws and lines of evolution act independently: a system first undergoes transformations along a particular line, and after this line is exhausted, switches to another line. The real picture is much more complex, because the historical development of any technological system is guided by intertwined interactions of several laws and lines of evolution.

For example, after a mono-system has evolved into a bi-system, the latter is usually characterized by rigid connections between its elements. While the system continues its advancement along the lines mono-bi-poly, it also starts developing flexible connections between the elements. An illustration of this is a catamaran whose hulls are connected together by stiff cross-links. Such a catamaran has poor maneuverability when passing through narrow waterways. A more advanced design uses cross-links that allow for changing the distance between the hulls, thus enhancing the navigational capabilities of the catamaran.

5.12   Summary

The foundation of TRIZ is a set of the laws of technological system evolution. These laws delineate the prevailing trends of the evolution of technological systems (i.e., products, technologies, and production processes); naturally, they are used both for identification and conceptual development of the next likely technological innovations. The predictive power of some of the laws is significantly reinforced by associated *lines of evolution* that specify stages of the system's evolution along a general direction. The use of these lines is rather direct. First, a technological system of interest is positioned on a line of evolution. Then the next evolutionary change, suggested by the line, is proposed for the system. This routine is repeated for each applicable line of evolution.

5.13   Exercises

5.1 Select any five sub-systems of a car. Apply the law of non-uniform evolution to each of these sub-systems and formulate system conflicts associated with future evolutions of these sub-systems.

5.2 (A) Give five examples of consumer products that can be characterized as:
- homogeneous bi-systems
- heterogeneous bi-systems
- partially convoluted bi-systems
- inverse bi-systems
- completely convoluted bi-systems.

(B) Give five examples of products from one industry that can be characterized as:
- homogeneous bi-systems
- heterogeneous bi-systems
- partially convoluted bi-systems
- inverse bi-systems
- completely convoluted bi-systems.

5.3 (A) Give one example of the evolution of a product along each of the following lines of evolution:
- Mono–bi–poly
- Transition to self-adaptive systems
- Transition to a micro-level
- Line of completion
- Shortening of energy flow path.

5.4 A conventional screwdriver with a straight or a Phillips blade can be considered a mono-system. A universal screwdriver with blades of different shapes (both straight and Phillips) and sizes, placed in its handle, is an example of the next evolutionary step – a homogeneous poly-system with shifted properties. Use the lines "mono–bi–poly" to develop at least one conceptual design of the next-generation screwdriver.

5.5 Single glass lenses can bring objects into sharp focus only at certain distances. Such lenses can be considered "stiff" systems. The single lens deficiency is eliminated in the two- and multi-element lens systems that are designed so that their focus can be varied by mechanically changing the axial spacing between their elements.

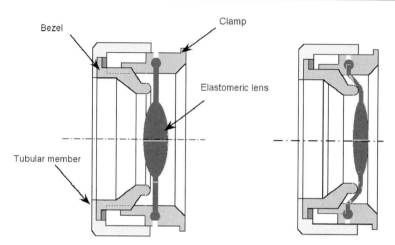

Fig. 5.43. Elastomeric lens system.

These lens systems are most commonly used in practice and they represent the "one-joint" and "multi-joint" stages of evolution respectively.

All conventional varifocal lens systems share common fundamental drawbacks: (a) considerable weight and space occupied by numerous glass elements, and (b) expensive and time-consuming manufacturing (i.e., grinding and polishing). These shortcomings are overcome in systems that use elastomeric materials and optical fluids instead of glass lenses.

A typical elastomeric variable focus lens system, Fig. 5.43, comprises a lens element formed by a transparent resilient material. Clockwise rotation of the bezel causes the tubular member to displace axially along the optical axis. The leading edge of the tubular member applies force to the elastomeric material, thus stretching it and changing the lens shape. This results in changing the focal length of the lens from a short one to a longer one.

Fig. 5.44 shows the correspondence between the universal line of increasing flexibility and the stages of the lens evolution delineated above. However, evolution of lens systems cannot stop at the "fluid" stage. Develop conceptual designs of next-generation varifocal lenses.

5.6 Do you see a potential application of the lines of chronokinematics to your company's products? If the answer is negative, please explain.

5.7 A typical reed switch (Fig. 5.45) has a pair of thin ferromagnetic pieces (contacts) placed inside a glass tube that is wrapped up in an electric coil. When there is no electric current in the coil, the pieces do not touch each other. An electric current in the coil creates a magnetic field that magnetizes the pieces, thus pressing them against each other and closing the outer circuit. Use the lines of chronokinematics to develop a more energy-efficient reed switch.

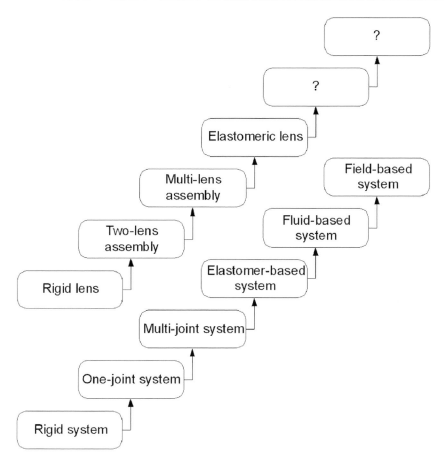

Fig. 5.44. Evolution of lenses vis-à-vis the line of transition to fluids and fields.

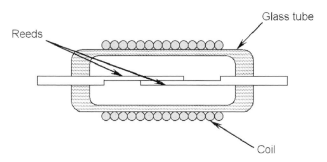

Fig. 5.45. Typical reed switch.

5.8 Apply any three lines of increasing flexibility to any consumer product of your choosing. Describe the resultant concepts and/or design challenges (system conflicts) that need to be overcome to realize these concepts.

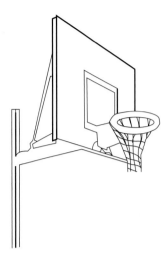

Fig. 5.46. Typical basketball system.

5.9 Apply the lines of mono–bi–poly to any consumer product of your choosing. Describe the resultant concepts and/or design challenges that need to be overcome to realize these concepts.

5.10 Select any three sub-systems (or components) of a car. Apply the line of completion and the line of transition to self-adaptive systems to each of these sub-systems (components). Describe the resultant concepts and/or design challenges that need to be overcome to realize these concepts.

5.11 Fig. 5.46 shows a typical playground basketball hoop system. Use the laws of completion, increasing flexibility, and transition to a higher-level system to develop conceptual designs of new generations of this system. Compare your concepts with those available in the US patent database (http://www.uspto.gov) and with already commercialized products.

6 Guiding technology evolution

The following is an outline of TechNav – a comprehensive process for the conceptual development of next-generation technologies and products based on the laws and lines of technological system evolution and business analysis (Fig. 6.1). The major points of this process are listed here.

(1) *Phase 1: analysis of the past and current system's evolution.* This stage is essentially about the analysis of an S-curve. The evolution of any technological system usually follows an S-shaped curve that reflects the dynamic of the system's benefit-to-cost ratio from the time of the system's inception (Fig. 6.2). An analysis of innovation activity allows for positioning the technology or product of interest on its S-curve, thus helping to define the technology (product) strategy.

(2) *Phase 2: determination of high-potential innovations.* At this phase, the laws and lines of evolution are used to identify strategic directions of the technology's evolution. In addition, the current innovation landscape (strategy) is analyzed and new ideas are developed.

(3) *Phase 3: concept development.* As a rule, transition from one stage of evolution to the next is accompanied by the development of system conflicts and other conceptual design problems that need to be resolved.

(4) *Phase 4: concept selection and technology plan.* In this phase, the developed concepts are evaluated against various engineering, economic and other criteria, and the best ones are selected for both short- and long-term testing and implementation.

6.1 Phase 1: analysis of the past and current system's evolution

The evolution of a technological system moves through four typical phases (Fig. 6.2). In the "infancy," or pre-market, phase, the rate of system development is relatively slow. In the next "rapid growth" phase, the system enters the market, and the pace of its development escalates exponentially. This is usually attributed to the practical implementation of the system, and the perfecting of its manufacturing processes. Eventually, the system's evolution slows down and stalls, or even starts degrading. These segments

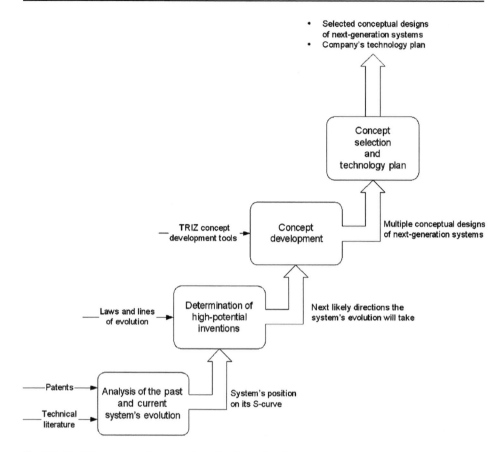

Fig. 6.1. TechNav process for strategic technology planning.

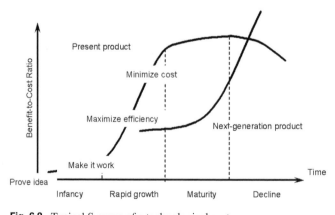

Fig. 6.2. Typical S-curve of a technological system.

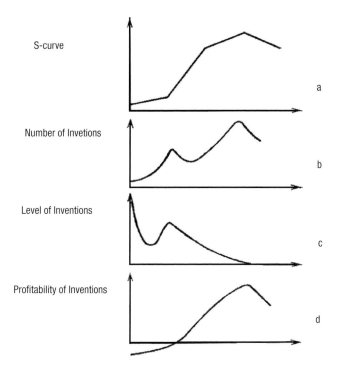

S-curve

Number of Invetions

Level of Inventions

Profitability of Inventions

a

b

c

d

Fig. 6.3. Correlation between innovation activity and S-curve.

of the system's "life curve" are typical for its "maturity." In some cases, the system undergoes a "renaissance," which can be sparked by the availability of new materials, new manufacturing technology, and/or by the development of new applications. While the present system approaches the end phase of its development, a new system holding the promise of a higher benefit-to-cost ratio is usually already waiting in the wings.

The length and slope of each segment on the system's "life curve" depends not only on technological, but also on economic and human factors. While common sense (hindsight!) suggests that a new system in Fig. 6.2 should begin its rapid development when the development of the present system begins to lose pace, it is frequently deferred by special interest groups that have a huge investment, job security, etc., associated with the old system.

Analysis performed by Altshuller demonstrated that inventive activity, associated with a technological system, closely correlates with the S-curve. Figure 6.3a is the S-curve for a system, while Fig. 6.3b is a typical plotting of "number of inventions as a function of time" for this system. Figure 6.3b has two peaks. The first peak occurs around the beginning of the wide implementation of the new system. The second peak occurs at the end of the "natural life" of the system, and relates to efforts (usually futile in the long run) to extend the system's life to compete with the new evolving system.

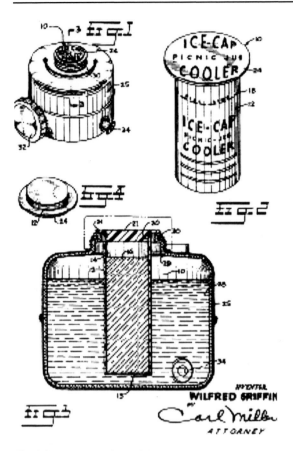

Fig. 6.4. Invention of Level 1.

Although most inventions usually advance the state of the art, the contribution of each individual invention can vary within a very broad range. To monitor the evolution of a technological system, it is important to study inventions of significant importance related to this system. Altshuller suggested dividing all inventions into five novelty levels:

6.1.1   Inventions of Level 1

These are slight modifications of existing systems. They do not resolve any system conflicts and are localized within a single sub-system. They are also contained completely within a narrow specialty of engineering, and are within reach of any practitioner.

Example 6.1
US Patent 3,059,452 was issued for the design of "Ice Cap Container for Picnic Jug Cooler". In this design (Fig. 6.4), a container of ice is inserted into the internal cavity of the cooler. Although there is a useful purpose in this design (neat appearance of

Fig. 6.5. Conventional dead centers allow external machining only.

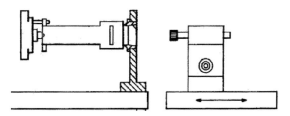

Fig. 6.6. Hollow dead center allows internal machining.

the cooler, easiness of replacing the container), the principle is well known and has numerous known applications.

6.1.2   Inventions of Level 2

These inventions resolve some system conflicts that may have already been resolved in other systems (e.g., a problem related to cars is solved by a technique developed for trucks).

Example 6.2

The precision machining of external surfaces of cylindrical parts (turning on lathes, grinding on OD grinders) is frequently performed when the part is supported between two stationary ("dead") or rotating ("live") accurately fabricated cones (centers) (Fig. 6.5). The dead centers are inserted into matching conical sockets in the part. These sockets are drilled, with special tools, into both ends of the part as its first machining step. The part is rotated by a driving pin, engaging with a collar clamped to the part. Such an arrangement, in which the same accurate reference surfaces are used both for rough and finish machining, results in both a high accuracy and concentricity of the cylindrical surfaces.

This arrangement, however, cannot be used for the boring or internal grinding of internal cylindrical surfaces, since there is no room for the passage of the tool inside the part. US Patent 5,259,156 describes a hollow dead center that allows for the passage of the cutting tool, while supporting the part (Fig. 6.6). This simple improvement resolves the system conflict and results in a better dimensional accuracy, especially concentricity, of machined parts.

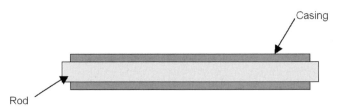

Fig. 6.7. Compensation of the rod's thermal expansion due to the Poisson's effect.

6.1.3   Inventions of Level 3

These inventions resolve system conflicts by an original approach within one discipline (e.g., mechanical engineering, chemical engineering, etc.). They also radically change at least one system's component.

Example 6.3
The length of a metal rod changes with a varying temperature due to thermal expansion. There is a need in precision machines and instruments to maintain the length constant. In some applications, where the use of alloys has reduced the coefficient of thermal expansion (CTE) (e.g., Invar whose CTE is about ten times lower than that of steel) is adequate. However, in some cases the requirements are so stringent that it is necessary to control the effects of temperature changes even in Invar.

   US Patent 3,528,206 suggests a design (Fig. 6.7), in which an annular tight-fitting outer metallic casing is shrink-fitted onto a rod. The shrink-fit causes radial squeezing/compression of the rod that is accompanied by axial elongation (Poisson's effect). The casing material's CTE is such that, with an increase of the ambient temperature, the casing expands outwards. This removes some of the radial compressive force on the rod; as a result, the rod shortens. If the materials and geometry of both the rod and the casing are properly selected, the rod's thermal expansion can be largely compensated.

6.1.4   Inventions of Level 4

These inventions give birth to new systems using interdisciplinary approaches. The developed concept can usually be applied to many other problems at the lower levels.

Example 6.4
While studying (in the 1940s) the erosion of electrodes due to sparking during their disconnection, Russian scientists Boris and Natalia Lazarenko suggested beneficially using this effect for the machining of hard materials and/or very intricate surfaces. Tools made of easily machinable materials, such as copper or graphite, are used to generate complex three-dimensional surfaces on hard-to-machine parts, such as hardened steel parts or parts made of super-hard tungsten carbide. Thus, the mechanical

Table 6.1. *Distribution of inventions by novelty level*

Level of invention	Criterion	Relative share
Level 1	A component intended for the task is used. No system conflicts are resolved.	32%
Level 2	Existing system is slightly modified. System conflicts are resolved by the transfer of a solution from a similar system.	45%
Level 3	System conflicts are resolved by radically changing or eliminating at least one principal system's component. Solution resides within one engineering discipline.	19%
Level 4	System conflicts are resolved and a new system is developed using interdisciplinary approaches.	<4%
Level 5	Resolving system conflicts results in a pioneering invention (often based on a recently discovered phenomenon).	<0.3%

cutting of materials was replaced by a direct controlled application of electrical energy in this breakthrough electro-discharge machining (EDM) technology.

6.1.5   Inventions of Level 5

These are pioneering inventions that are usually based on recently discovered phenomena. They often result in the creation of a new engineering discipline (invention of the airplane; invention of radio transmission; invention of computers; invention of lasers; etc.).

Not surprisingly, the distribution of the inventions by novelty levels is quite non-uniform (Table 6.1).

The levels of inventions at different stages of a system's life are shown in Fig. 6.3c. In the beginning, high-level inventions (4 or 5) create the basis for a new system. The levels gradually decrease and then rise (the peak on the curve) when many problems associated with manufacturing and marketing must be solved. After this peak, the low-level inventions dominate again.

Invention activity analysis should also be performed for competitors' products and technologies.

Managers need to understand the dynamics of these life curves. This can help them answer an important question: "Is it worthwhile to improve the existing technology (product), or is it more beneficial to develop a new one?"

It is usually possible to collect information about a system's evolution status and to plot the progress of its principal parameter (e.g., speed, productivity, power, accuracy, etc.). Three typical cases may exist.

(1) The system is still in its infancy stage. At this stage, one can expect major breakthroughs since there is a need to make a *new product* out of the *new idea*. To this end, the *new idea* (i.e., an invention of Level 4 or 5) has to be supported by lower-level inventions. However, the journey to the next, rapid growth stage is usually held back by the present dominant system. The latter deters the development of

the new competing system. Both technical and non-technical factors influence this evolutionary process.

(2) If the system is in the rapid growth stage, forecasting should determine its physical limits based on objective factors (e.g., strength of materials, heat-producing capacity of fuels, various barriers, such as speed of signal propagation in integrated circuits, etc.).

(3) The system has matured or is in decline. In this case, the forecasting process is a search for a new system that would succeed the present one.

In all these situations, managers can embrace one of the two approaches: leapfrog to a new system, or concentrate on incremental improvements to the old system. The company has to make the choice; the role of TRIZ is to present them with the alternatives.

The authors participated in the development of the next-generation coffee brewer undertaken by a leading manufacturer of this product. The coffee brewing industry is overcrowded with numerous rival companies, and the competition is very tough. Analysis of 1239 US, European, and Japanese patents, issued from 1975 to 2000, showed that most of the inventions in the coffee brewing field were of Levels 1 and 2, i.e., minor improvements not significantly advancing the state-of-the-art. The analysis helped to objectively position the company's product (as well as those of its competitors) on its S-curve. It became clear that the existing coffee brewing technology had long matured and approached its limits. In order to attain a robust competitive advantage, the company had no other real choice than to come up with technological breakthroughs based on new scientific principles. This analysis triggered a search for radical innovations. Several breakthrough solutions were developed and tested. The worldwide patenting of some concepts has been undertaken (US Patent No. 6,805,041; G.B. 2,366,510).

Summing up, Phase 1 – analysis of the past and current system's evolution – takes the following steps.

(1) Identify major performance parameters of the system.

(2) Collect patents associated with the identified parameters.

(3) Determine levels of inventions described in the patents.

(4) Determine the position of the technology on its S-curve by plotting both the average level of inventions and the number of inventions vs. time.

(5) Determine the levels of inventions associated with the company's technology.

When selecting patents for the analysis, it is helpful to take into account the following rules of thumb.

(1) Select patents with a high citation frequency (≥ 3).

(2) Low citation frequency is not always a sign of an inferior technology. The technology can be new, or may need a "booster" technology to accelerate its development. Due to a higher rate of evolution in hi-tech industries, "high-tech" patents may not adequately reflect the state-of-the-art. Use technical and scientific journals, corporate R&D publications, etc. to determine the "cutting edge".

(3) Real-life plots are not smooth (see, for example, Fig. 6.8).

Number of patents

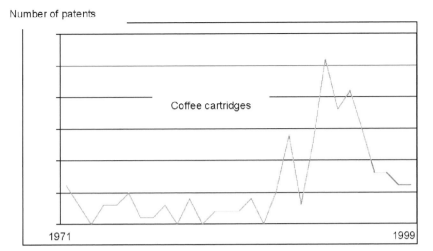

1971   1999

Fig. 6.8. Example of a real-life technology assessment curve built for coffee cartridges used in coffee brewers.

Having completed the S-curve analysis, the managers face two principal questions.
• What changes should be made to my product (technology) to move it up its S-curve?
• What new products (technologies) may potentially replace my product?
The next two phases of the TechNav process help answer these questions.

6.2   Phase 2: determination of high-potential innovations

At this phase, the laws and lines of technological system evolution are used. The first step is to use the concept of *domination of higher-level systems*: *evolutionary trends of a system determine the evolution directions of its components*. This means, in particular, that a problem at the system's level may be a symptom of a problem at a higher hierarchical level.

Example 6.5
An electronics equipment manufacturer was planning to enter a new market by developing a next-generation device, let's call it a controller, used in fluid analyzers. These analyzers are employed in some manufacturing processes and biomedical procedures. A conventional fluid analysis involves two steps: first, a sample of the fluid is taken from the pipeline into a special receptacle; then, the receptacle is placed into a fluid analyzer.

A preliminary marketing analysis, conducted by the company, showed that prospective customers needed a controller that would be free of certain disadvantages that were characteristic of the commercially available controllers.

One of the lines of evolution shows that, in the course of their historical development, system components performing auxiliary functions gradually merge with components that carry main functions (the law of transition to a higher-level system). Eventually, the auxiliary components may entirely disappear by delegating their functions to the main components. In the case of the fluid analysis, the pipelines are main components, while the fluid analyzers are auxiliary ones. This suggested a future integration of the fluid analyzers with the pipelines. Accordingly, the controllers' evolution would depend on this integration. A simple engineering review of this option showed that inline fluid analyzers would not need the function performed by the controller.

As a result of this analysis, the company conducted new marketing research. It turned out that the process of integration of the fluid analyzers into pipelines had already started, and that some potential customers had shown interest in it. The company modified its technology plan and began developing components for the inline fluid analyzers.

Frequently, an application of the laws of technological system evolution results in conclusions that challenge the company's technology strategy. The management should then modify this strategy in light of the new findings, or stick to the old one, but know that they are gambling with the logic of technology evolution.

The next activity is the imposition of the lines of technological system evolution on the system of interest.

6.3   Phase 3: concept development

Application of the laws and lines of evolution to a particular system may result in three outcomes:

Outcome 1: both the physical principle of the functioning of the next-generation system, and all of the system's components are identified.

This scenario is less frequent, because the laws and lines of evolution may be too broadly formulated to be employed for the development of specific conceptual designs. It is also the least interesting from an invention perspective (most of the creative work has already been done).

Outcome 2: the physical principle of the system's functioning and some of the system's components are identified. Identification of the missing components requires the resolution of certain system conflicts.

Example 6.6
Conventional computers are stand-alone devices. While they may be moved from place to place, they are not meant – even laptops or tablet PCs – to be constantly carried around by the user. However, many people would prefer constant access to computers.

The law of transition to higher-level systems, when applied to personal computers, points at the inevitable integration of the latter with the user, and first of all with clothes.

Recently, wearable computers have indeed become available. Presently, these computers feature low-performance processors yet, as the technology improves, more powerful processors will be used. High-performance integrated circuits usually generate a substantial amount of heat. Since this heat may adversely affect both the computer and the user, the integrated circuit chips must be cooled, and the heat must be somehow dissipated. However, conventional cooling devices, such as fans, heat sinks, heat pipes and the like, are too bulky and/or power-hungry to be practical for wearable computers. Thus, for these computers to become an everyday item, a system conflict between portability (miniaturization) and heat generation will have to be resolved.

Example 6.7

On page 174 we briefly touched on our participation in the development of next-generation coffee brewers. The following case story was a part of that project.

Most commercially available coffee brewers serve one type of drink: regular coffee (sometimes of different strengths). An application of the law of transition to a higher-level system to such a single-drink machine leads to, among other ideas, the concept of a coffee brewer that serves other types of coffee drinks, for example, espresso. The development of such a product confronts a specific challenge.

Many modern coffee brewers make drinks from flexible thermoplastic film sachets or cartridges containing a single portion of the coffee powder (Fig. 6.9). The sachet has an injection nozzle at the top and a seal at the bottom. A paper filter is placed in a lower part of the sachet. In use, hot water is injected through the nozzle into the sachet. The drink is brewed inside the sachet; the seal at the bottom is opened at the end of the brewing to allow the drink to flow into the cup. This brewing method provides excellent quality, fresh coffee. However, the thin sachets cannot withstand the high pressure (9–15 bar) required to brew espresso coffee. Thick sachets reinforced with a layer of metal foil are strong enough, but recycling them is more difficult (which contributes to landfill waste).

A brewing chamber with rigid walls could prevent the bursting of the espresso sachet. However, such a chamber could not accommodate sachets of different sizes.

Another dilemma: the brewing chamber of the coffee machine must rigidly constrain the espresso sachet to generate and maintain high water pressure for preparing high quality espresso; on the other hand, the brewing chamber must provide for relatively low pressure "flooding" of the "regular" coffee sachet to prepare that high quality drink.

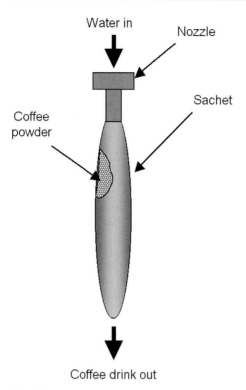

Water in

Nozzle

Sachet

Coffee powder

Coffee drink out

Fig. 6.9. Coffee sachet.

These demands can be expressed as a physical contradiction: the brewing chamber walls must be both rigid and flexible. This statement clearly calls for a solution providing a separation of the contradictory properties in time: the walls are rigid during brewing espresso and flexible when preparing a regular coffee drink.

Figure 6.10 shows a schematic cross-section of one of the embodiments of this approach: a brewing chamber with elastomeric (dynamic) clamps. These clamps enable the size and shape of the brewing cavity to vary simply by varying the closing force applied to the rigid base members. This provides a greatly increased flexibility and control over the brewing process. The coffee brewer equipped with these clamps can brew espresso coffee by tightly clamping the sachet and then injecting water at high pressure. The conformability of the elastomeric clamps supports the walls of the sachet against bursting. The elastomeric clamps can be adapted for brewing filter coffee simply by compressing the sachet less tightly and by injecting a larger amount of water relative to the amount of coffee in the sachet.

Other options of the dynamic clamps are described in GB Patent 2,366,511.

Outcome 3: high-potential direction for the system's evolution is identified. Both the physics and the structure of the system are yet to be specified.

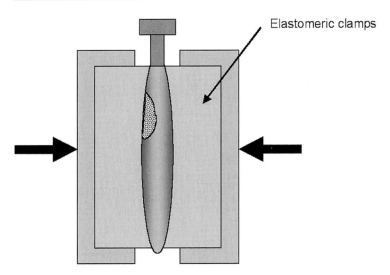

Elastomeric clamps

Fig. 6.10. Flexible sachet clamps (GB Patent 2,366,511).

Example 6.8

As a hypothetical example, suppose we apply the law of transition to micro-level to the conventional punch-and-die metal sheet forming process. This process employs mechanical phenomena to shape the workpiece. The law of transition to micro-level points toward replacing the mechanical sheet deformation process with one based on electromagnetic phenomena. The law, however, does not specify what these phenomena might be. Of course, we are free to speculate about a new physics of metal forming, but a systematic approach to an identification of this physics would be desirable.

We can also apply another law of evolution to the original metal sheet forming process, for example, the law of shortening the energy flow path. In that case, we would be advised to eliminate the intermediary subsystems (electric motor, transmission, and punch) and apply the deformation force directly to the metal sheet. Yet again, the law itself does not contain any information on the specific physical effects that might be used to realize such a process.

It is clear that sufield analysis (see Chapter 3) should be applied in this situation to identify the most appropriate physical effects.

Example 6.9

Peristaltic pumping is a form of fluid transport that occurs when a progressive wave of area contraction or expansion propagates along the length of a flexible tube containing a liquid or gas. Peristaltic action occurs in the gastrointestinal tract, the bile duct, and other ducts of the body.

In engineering, peristalsis is used in roller and finger pumps. These pumps are used to transport slurries, foods, corrosive fluids, or blood, whenever it is desirable to isolate

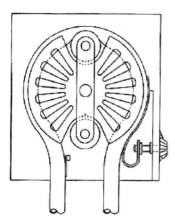

Fig. 6.11. Conventional peristaltic pump.

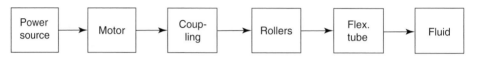

Fig. 6.12. Chain of energy transmission/conversion in the conventional peristaltic pump.

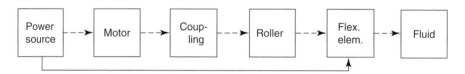

Fig. 6.13. Energy transmission/conversion in a novel peristaltic pump.

the transported fluid from the mechanical parts of the pump. In these devices, the tube is usually compressed by rotating rollers or fingers (Fig. 6.11).

Application of the laws of technological system evolution
Major components of a peristaltic pump make up a long chain of energy transformation (Fig. 6.12). This calls for invoking the law of shortening the energy flow path. The system's energy efficiency would be much higher, and the pump design would be much simpler, if an electromagnetic power were directly applied to a flexing element (Fig. 6.13). The latter cannot be a conventional flexible tube, because that is unreceptive to an electromagnetic field.

Concept development
This requirement easily translates into a typical sufield structure (Fig. 6.14). Now, by pairing various substances with an electric and magnetic field, one can generate an array of conceptual designs for the next-generation peristaltic pump.

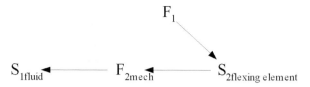

Fig. 6.14. Sufield model of a novel peristaltic pump.

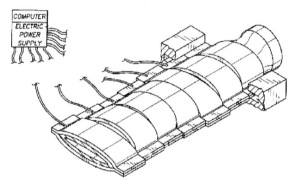

Fig. 6.15. Electrostatic peristaltic pump (US Patent No. 5,705,018).

Figure 6.15 shows one of possible designs employing the pair $F_{electric} \rightarrow S_{2electroconductive}$. In this design, described in US Patent No. 5,705,018, a channel, formed in a semiconductor substrate, is covered with a flexible electroconductive membrane. Insulative barriers in the channel separate parallel electroconductive strips. Sequential application of a voltage to the electroconductive strips pulls the membrane inside the channel, thus generating a wave that creates a pumping action.

One can also conceive pump designs based on the pairs $F_{magnetic} \rightarrow S_{2ferromagnetic}$ (e.g., see US Patent No. 6,074,179), or $F_{electric} \rightarrow S_{2ceramics}$ (see US Patent No. 6,074,178), and some others.

The described peristaltic pumps are more energy efficient, much less complex and, therefore, more reliable than the conventional designs. Their small size makes it easy to incorporate them into various other systems, such as heat exchangers for printed circuit boards and semiconductor components, fluid analyzers, or to use them as micro-pumps in portable blood- and other bodily-fluid-moving devices.

6.4 Phase 4: concept selection and technology plan

In this phase, the developed concepts are evaluated against various engineering, economical and other constraints and the best ones are selected for both short- and long-term testing and implementation.

A special effort is made to maximally mitigate the risks of new technologies. To achieve this, the concepts are ranked as incremental and radical changes. The risk associated with the concepts is determined by considering a variety of criteria, such as:
- was the proposed technology tested?
- was a physical (engineering) analysis of this technology performed?
- is the technology at its theoretical limit? Is the technology accessible to the company?

Based on experiences with analogous technologies, estimates of the required time to develop and prepare new technologies for market are made.

6.4.1   Preparing the technology plan

Most product companies use technology planning to outline the technologies they plan to invest in, the amounts they will invest, and the timing of those investments.

A technology plan consists of a series of investments made in three key areas.
(1) Product planning.
(2) Competitive analysis (present or potential know-how being developed by competitors, which is different from the company's technology).
(3) Advanced technology (technology where a design is not yet defined, but thought to be the potentially key to future products or subsystems).

Management must decide how to allocate resources between these needs. The timing of future technical evolution is critical to this resource allocation. TechNav can help plan investments in a way compatible with the likely evolution of technology. For each product and key subsystem in a company's product line, a TechNav evolution analysis can be made. At each step in the evolution, the key technologies necessary to achieve the evolution can be identified. Judgment can be made as to which branch of the fork (Fig. 6.16) is most likely to be taken, first by looking at the technologies necessary for each step to occur. For example, in Fig. 6.16, if all of the technologies needed for option A to evolve are available now, and the technologies necessary for option B to be realized are not, then A will happen first. The lead time that A will have over B will be the time necessary to develop and test technology B plus the development cycle time inherent to the specific industry in question. This cycle time is dependent on the need for testing, government approvals, etc., as well as on engineering development time.

To apply this method to technology planning, a three-step approach is suggested:
(1) Product line evolution; (2) Advanced technology; (3) Competitive analysis.

By developing a comprehensive technology plan in this way, the technology risks inherent in the business will be identified, and investment can be planned in a systematic way to manage the risks. The improvements in investment performance will come from the fact that the forecasts made for the technology evolution of both the product line and the competitors' products will have been done systematically using TechNav.

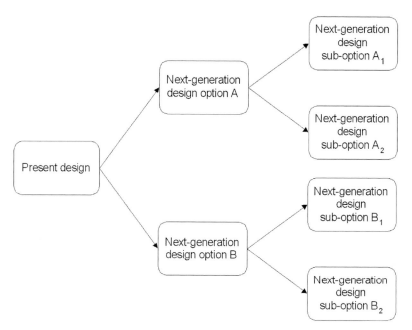

Fig. 6.16. The time line of technology evolution.

The above three steps have the following content.

Step 1: Product line evolution

- Place all the developed alternatives on the time line chart using normal industry cycle times.
- Allocate resources over the time chart.

Step 2: Competitive analysis

- Evaluate the need for the company's future products to compete with the competitors' predicted products.

Step 3: Advanced technology plan

- Make a list of all the technologies identified in the product line evolution plan.
- Prioritize them by:
  - (a) their importance to the product evolution, and
  - (b) their position on the time line chart.
- Justify the plan by showing the logic behind it.

These procedures result in a technology plan based on the principles of technological evolution. It is compatible with the patterns of evolution that have been shown to apply successfully to past developments. Is it a foolproof forecast? No, but it will substantially reduce the risks inherent in technical investments by being accurate most of the time. Combining these results with good technical judgment will vastly improve investment performance.

Table 6.2. *Next generation vehicle requirements*

Load	1000–2000 lbs
Range	250–300 miles
Top speed	75–100 m.p.h.
Time to "re-energize"	< 10 min
Efficiency	40–80 m.p.g.

Example 6.10

The following is a digest of the study performed by Dr. Norman Bodine of The TRIZ Group, LLC in 1999–2000 (Fey *et al.*, 2001; http://eyeforfuelcells.com/ReportDisplay2.asp?ReportID=664).

Controversy surrounds vehicle powertrains being developed by today's automakers. The US Government spent millions of taxpayer dollars studying the problem in order to create a set of standards needed for a next generation vehicle (Table 6.2).

In addition, there are safety and emission requirements specified by government and standards of reliability and comfort that increase constantly. Emission targets are 'near zero' – at a minimum 50% of today's levels. No sacrifice in comfort, safety or reliability should result when meeting these standards. Emission requirements are always subject to the whims of government regulation and are therefore subject to change. Efficiency is also regulated, but indirectly and not severely.

A similar list of requirements for today's powertrain would show all of the above requirements with a fuel efficiency of 20–40 miles per gallon. Market driven requirements for today's powertrains have been in place, and have evolved, for over 100 years. In summary, the objectives are to double efficiency, halve emissions and meet all other requirements.

A quick study of vehicle performance over all speeds and accelerations identifies a clear system conflict. When accelerating, an internal combustion (IC) engine is very inefficient and polluting. When cruising, however, it is more than adequate. In fact, most vehicles today could achieve the required efficiency at their cruising times with a much smaller internal combustion engine than is presently used. This conflict can be naturally separated in time. A different engine is needed when driving a high torque load (accelerating or going uphill).

With TRIZ, resolving system conflicts should, firstly, involve existing resources. Here an additional existing technology will be used, so that the conflict is eliminated. What is needed here is to add an existing machine, efficient and clean, when driving a high torque load.

Fortunately, there exists a machine, even older than the internal combustion engine that can do this job: the electric motor. So a TechNav analysis would suggest that option A in Fig. 6.16 is a hybrid powertrain consisting of a smaller IC engine used for cruising regimes and an electric motor used for acceleration events. This type of powertrain is likely to be the next generation powertrain, according to the TechNav forecast.

Secondary system conflicts naturally result from this proposal, and each of them can be approached using a similar analysis. The need for an electronic control system to manage the sharing of power between the electric motor and the IC engine is a good example. In fact, before some recently developed electronics became available, the control system would have been hard to realize, and the evolutionary timetable would have been pushed further into the future. Today, these technologies exist and are proven, so the analysis of the secondary system conflicts brings the timetable for the hybrid powertrain into the present. It can all be done now!

What about alternative technologies? Is there a logical sub-option to this solution, A-1 — another approach that can be developed with the rules of TechNav? It is difficult, because resolving the system conflict using other known approaches also requires the electric motor as the driving mechanism.

So TechNav predicts that these alternatives, such as fuel cells, would most likely occur as the next steps in the evolution, after A-1. They would evolve as efforts to improve the performance of the hybrid are made, with solutions to system conflicts arising later along the evolutionary path.

What about a vehicle powered by batteries and an electric motor? Today, the requirements for range distance and re-energizing time are way beyond the best present battery technology.

What about fuel cells? We have already identified them as a possible sub-option A-2. Could they happen quickly? Could the development time between A and A-1 be short? Fuel cells are not a new technology. They have been used as power sources in space for years. The only commercial application to date has been to provide electric power to very remote sites where transmission lines are extremely difficult to construct and costly power is acceptable. Fuel cells are not competitive in the market for standby generators, now dominated by internal combustion engines (the same or similar to those in vehicles) and driving electric generators (reverse motors). In this market, where safety issues are more manageable than in mobile vehicle applications, fuel cells might arise as a result of environmental regulations on emissions from standby generators. As yet, this possibility has not been realized. This leads to a conclusion that the time between A-1 and A-2 will likely depend on the time it takes to invent new fuel-cell technologies, government regulation and the business cycle of the automotive industry.

Let's now discuss some business implications of this TechNav forecast for the automotive powertrain. Given the lead-time for planning, tooling, production and market introduction of a new vehicle technology (currently about 3–4 years), the company that invests first in the right technology will have a significant lead on the competition.

The TechNav analysis suggests that the investment made by General Motors in the E-car (an all battery-powered vehicle) was premature. Some of the know-how can be adapted to a hybrid of the type the TechNav analysis suggests will come next, but much cannot. In any case, a TechNav analysis by GM management prior to the decision to design, tool, and produce the E-car might have saved GM a billion dollars or more. In addition, much of the valuable time lost on developing the technologies needed for the hybrid could have been saved.

Toyota and Honda, however, arrived at the same conclusion as the TechNav analysis. They have completed the design and production phases of hybrid vehicles, which have been on the market in Japan for 4 years and in the US for 2 years.

All the other competitors have been on the sidelines, presumably watching to see what happens. This is the typical attitude of many top level managers of automotive technology, "I don't want to be the first, I want to be the second." According to their press releases, they have been investing in many of the technologies mentioned above: fuel cells, alternative fuels, batteries and hybrids of various types, including turbines to replace internal combustion engines. If they wait for the market to sort out the next technology, they will lose even more time.

With the time for development and production, the lead that Honda and Toyota have in hybrids is 4 years and climbing. Meanwhile, they gain valuable field experience with their hybrids that they can use to improve their designs. They do this while volumes are low, as is the cost of learning. The longer other competitors wait, the more likely it becomes that they will do their learning when volumes are high on the S-curve and when mistakes are costly, both financially and in customer satisfaction.

Transition to the hybrid powertrain reflects one of the prevailing lines of evolution: "old technology $\rightarrow$ old technology $+$ immature new technology $\rightarrow$ refined new technology". History provides a plethora of examples of hybrid technologies: steam engines on sail ships, radial tires with a diagonal carcass, radios using both vacuum tubes and transistors, and many others.

One of the advantages of hybrids is that they give an opportunity to discover problems associated with the new technology while introducing it into the market. Naturally, the buffer of a hybrid technology can withstand the assault of the new technology for only so long; sooner or later it will be overcome. The management should use this time to adequately prepare the company for a full transition to the new technology.

All of this illustrates the critical importance of technology planning, and how it can have a huge impact on the success of a product-manufacturing business. It shows how vital it is that investments be made in the right technology – and at the right time.

An analysis of the type shown in this example can be done for any product where technology is an important factor. It provides management with something they have needed for a long time: a method to estimate how technology in their business will evolve and to guide them in their investment decisions, thus dramatically reducing the risks.

6.5   Summary

Technology is pivotal to competitive advantage, but technology investments are risky. To remove the guesswork from pinning down the next winning technology, and to provide scientific justification for investment decisions, technology managers can use the TechNav process to identify and develop the concepts of the next likely technological innovations.

The TechNav process includes four phases: Phase 1: analysis of the past and current system's evolution; Phase 2: determination of high-potential inventions; Phase 3: concept development; and Phase 4: concept selection and technology plan.

In Phase 1 the analyzed system is positioned on its S-curve. This helps technology managers and business leaders decide which path to take: to keep producing incremental improvements, or to leapfrog to a new S-curve. In Phase 2, the laws and lines of technological system evolution are used to identify the next, most likely directions that the system evolution will take. Specific conceptual designs that go along with these directions are developed at Phase 3. Selection of the most promising concepts is done at Phase 4.

Appendix 1

Genrikh Altshuller – the creator of TRIZ

Genrikh Saulovich Altshuller was born on October 15, 1926 in Tashkent, Uzbekistan (formerly the USSR) to a family of journalists. A few years later, the family moved to Baku, Azerbaijan (formerly the USSR).

Altshuller was awarded his first author's certificate (the Soviet-era equivalent of a patent) for a diving gear when he was in the tenth grade. While in high school, he also built and tested a boat with a jet engine that used acetylene as fuel.

After graduating from high school with honors, Altshuller was admitted to the Azerbaijan Industrial Institute. In 1944, with World War II still raging, he enlisted in the army. Although he was trained as a fighter pilot, the war was over before he had a chance to participate in combat. In 1946, Altshuller was assigned to the Commission on Innovation of the Caspian Navy Flotilla,[1] where he continued to invent in various fields of technology.

Altshuller was 20 when he began the research that would later become TRIZ. He decided to develop a method for systematic invention that could transform engineering creativity from magic, attainable by few, into a logical discipline available to many more. He conjectured that the key secrets of inventiveness should not be sought inside the inventors' minds, but rather in the logic of the inventions themselves. He realized that multiple industries and technologies used the same inventive principles. Altshuller

---

[1] The flotilla was headquartered in the city of Baku.

reasoned that if it was possible to extract these principles, than the pace of innovation would greatly accelerate. In search of such principles, Altshuller began studying patents and the histories of inventions. Between 1946 and 1948, he made these key discoveries: (1) a breakthrough invention is the result of overcoming a system conflict, and (2) technological systems evolve toward increasing ideality. He also proposed the initial formulations of the law of transition to a higher-level system and the law of harmonization of rhythms.

In 1948, Altshuller and his schoolmate and associate, Rafail Shapiro, wrote a letter to the Soviet dictator, Josef Stalin. The letter stated that the country was in ruins after the war, and that the resources needed for its recovery were scarce. To help the nation, the authors suggested using TRIZ. The reply came in 1949: Altshuller and Shapiro were arrested, interrogated and sentenced to 25 years in the notorious Vorkuta labor camp, above the Arctic Circle.

In the camp's coalmine, Altshuller toiled along with many representatives of the academic and industrial elite, who were slowly dying in the camp's brutal conditions. He realized that in order to survive, not only physically, but also spiritually and mentally, he had to continue his education and research. Altshuller opened a "One-Student University." Every night, in the barracks, former university professors taught him physics, mathematics, art history, literature and foreign languages. These lessons taught Altshuller a great deal, and allowed the academics to endure far longer than they would have without him.

In 1953, Stalin died. In 1955, Altshuller returned to Baku where he resided until 1990. A year later, the first article on TRIZ was published (Altshuller and Shapiro, 1956).

Since it was virtually impossible for a recent political prisoner to find a permanent job, Altshuller decided to make a living by writing science fiction. His first story, *Icarus and Dedalus*, was published in 1957. Writing under the pen-name Altov, he quickly became a popular author; his sci-fi stories and novels were translated into many languages (see, e.g., Altov and Zhuravleva,[2] *Ballad of the Stars,* 1982, New York, McMillan). The typical protagonist in the stories is a creator of some groundbreaking invention, living in our time. The author closely examines the philosophical and societal implications of the protagonist's inventions. Altshuller's stories are packed full of brilliant sci-fi ideas, some of them rather realistic. For example, in his *Donkey and Axiom*, published in 1965, he suggested that a light beam generated by an Earth-bound laser might propel future rockets. This beam would not only supply energy to the spacecraft, but would also carry information from the Earth to the astronauts. The concept of laser-propelled spacecraft has been recently discussed in scientific literature (see, e.g., Landis, 1989), and a prototype of such a vehicle is being developed by Lighcraft Technologies, Inc.

---

[2] Valentina Zhuravleva (1933–2004), Altshuller's wife, co-author, and long-time associate, was also a prominent sci-fi author.

In parallel with the development of TRIZ, in 1964, Altshuller began research into the mechanisms of generating new sci-fi ideas. This project culminated in the mid-1970s with *The Register of Sci-Fi Ideas and Situations*,[3] which contains about 10 000 ideas catalogued into classes, sub-classes, groups, sub-groups, etc. Analysis of this science fiction's "patent database" helped Altshuller develop many techniques used in his courses on creative imagination development.

Altshuller always worked from home. Nevertheless, from 1959 to 1985 he criss-crossed the Soviet Union holding dozens of workshops and seminars on TRIZ. The duration of many of these workshops was 2 to 3 weeks. In 1971, he founded the Public Institute for Inventive Creativity, which became the first center of TRIZ learning in the world. He also helped to organize local TRIZ schools all over the former USSR. By the end of the 1980s, over 500 schools existed.

From 1974 to 1986, Altshuller published dozens of tutorials on TRIZ in a national children's weekly magazine. The tutorials contained problems presented in an easy-to-follow, entertaining way. The readers – mostly middle- and high-school students – sent him letters with solutions to the problems. Through the analysis of about half a million of these letters, Altshuller wrote a bestselling book *And Suddenly the Inventor Appeared* in 1984.

Altshuller did not see TRIZ only as a system of powerful concept generation tools. For him, it was a means of developing skills for what he called *strong thinking* (this is a literal translation from Russian; another possible version would be *analytical independent thinking*).

Altshuller held the view that the most significant revolutions are caused by new powerful ideas. He also maintained that the well-being (both ethical and economical) of a society depends largely on the proportion of creative individuals in that society. A creative individual, according to Altshuller, pursues a major noble goal (some examples: Allen Bombard, Albert Schweitzer and Albert Einstein). To achieve this goal, the creative individual must be able to think innovatively, i.e., analytically, holistically and independently.

Altshuller also believed that many of the acute problems that humanity faces, and will inevitably confront in the future, might be eliminated (or prevented) but for our inability to think logically and independently. Humanity's only means of understanding and changing reality is reason. We cannot survive unless we fully develop and employ our intellectual power to achieve ethical and material goals that can assure the continual evolution of the human race. Consequently, Altshuller considered TRIZ to be a prototype for the future universal method of developing *strong thinking*.

To help raise creative individuals, in 1984 Altshuller and his student and associate, Igor Vertkin, started a new project – they analyzed hundreds of the biographies of

---

[3] This work is available in Russian only, at http://www.altshuller.ru/rtv/sf-register.asp.

innovators in science, technology, art, religion, politics, etc. The result was the monumental *Life Strategy of a Creative Person* (Altshuller and Vertkin, 1994).

In 1989, Altshuller became the President of the International TRIZ Association, founded by his former students. In 1990, he and his family moved to Petrozavodsk, Russia, where he passed away on September 24, 1998, due to complications from Parkinson's disease.

Altshuller's literary legacy is vast: 20 books, about 400 papers, and thousands of letters. His other legacy is not easy to quantify, but might be even more significant: thousands of people all over the world who, for the first time in their lives, have (and will have) experienced one of the most potent human emotions – the joy of creativity.

Appendix 2

System conflict matrix and inventive principles

The analysis of thousands of inventions by Altshuller resulted in the formulation of *typical system conflicts*, such as productivity vs. accuracy, reliability vs. complexity, shape vs. speed, etc. It was discovered that, despite the immense diversity of technological systems and the even greater diversity of inventive challenges, there are only about 1250 typical system conflicts.

Typical conflicts can be overcome by a relatively small number of *inventive principles*. There are 40 inventive principles for overcoming system conflicts, and each of them may contain a few sub-principles, totaling up to 100. The inventive principles and typical system conflicts are combined in a *system conflict matrix* (or the *Matrix*, for short). The Matrix contains attributes that are the most common for all technological systems (e.g., weight, strength, length, surface area, power, ease of usage, etc.). These attributes are arranged in rows and columns.

Suppose there is a need to improve some attribute of the system. If this improvement causes the worsening of another attribute (i.e., a system conflict develops), the intersection of the row and column corresponding to the conflicting attributes will show the potentially useful inventive principles (Fig. A2.1). The application of one of these inventive principles, or their combination, has a high probability of solving the problem.

The inventive principles are formulated in a general way. If, for example, the Matrix recommends the inventive principle *flexibility*, it means that the solution to the problem relates somehow to changing the degree of flexibility or adaptability of the technological system being modified. The Matrix does not solve the problem, it only suggests the most promising directions to search for a solution. The problem solver has to interpret these suggestions, and find a way to apply them to the problem at hand.

When using the Matrix, one can follow a simple procedure.

(1) In the Matrix, select an attribute that needs to be improved.
(2) Answer the following questions.
    (a) *How is it possible to improve this attribute using conventional means?*
    (b) *Which attribute would worsen, if conventional means were used?*
(3) Select an attribute in the Matrix corresponding to Step 2b.
(4) Using the Matrix, identify the inventive principles for overcoming the system conflict.

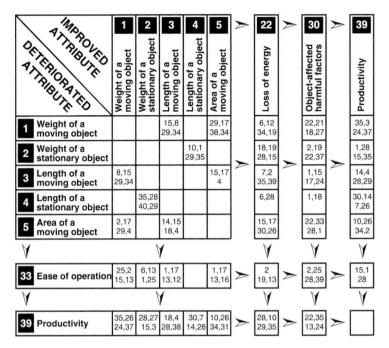

IMPROVED ATTRIBUTE / DETERIORATED ATTRIBUTE	**1** Weight of a moving object	**2** Weight of a stationary object	**3** Length of a moving object	**4** Length of a stationary object	**5** Area of a moving object	**22** Loss of energy	**30** Object-affected harmful factors	**39** Productivity
1 Weight of a moving object			15,8 29,34		29,17 38,34	6,12 34,19	22,21 18,27	35,3 24,37
2 Weight of a stationary object				10,1 29,35		18,19 28,15	2,19 22,37	1,28 15,35
3 Length of a moving object	8,15 29,34				15,17 4	7,2 35,39	1,15 17,24	14,4 28,29
4 Length of a stationary object		35,28 40,29				6,28	1,18	30,14 7,26
5 Area of a moving object	2,17 29,4		14,15 18,4			15,17 30,26	22,33 28,1	10,26 34,2
33 Ease of operation	25,2 15,13	6,13 1,25	1,17 13,12		1,17 13,16	2 19,13	2,25 28,39	15,1 28
39 Productivity	35,26 24,37	28,27 15,3	18,4 28,38	30,7 14,26	10,26 34,31	28,10 29,35	22,35 13,24	

Fig. A2.1. Fragment of the Matrix.

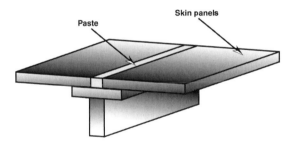

Fig. A2.2. Paste fills the gap between the skin panels.

(5) Try to identify another attribute, associated with the problem that might be improved.

(6) Repeat Steps 2 to 4 in relation to the attribute identified at Step 5.

Example A2.1

Thermal stresses in aircraft skin structure are undesirable. Compensation for thermal deformations is usually provided by small gaps left between aircraft skin panels (Fig. A2.2). To aerodynamically smooth the surface, a polymer-based sealant (paste) is used to close the gaps. Filling the gaps with the paste is a very labor-intensive and time-consuming process.

The disadvantages of this situation are associated, to various extents, with the sealant. What is an ideal sealant? Obviously, it is a sealant that is absent. Such a sealant does

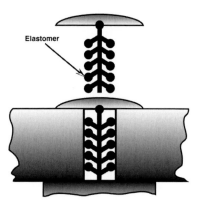

Fig. A2.3. Metal cap with an elastomeric retainer allows for the thermal expansion and reduces the drag.

not need to be placed between the panels, but at the same time, degrades aerodynamic characteristics of the skin. In other words, the attribute "ease of operation" causes a worsening of the attribute "object-affected harmful factors." For this pair of conflicting attributes, the Matrix advises one to examine the applicability of the inventive principles *separation, self-service, non-mechanical changes* and *neutral environment*. If the system conflict is formulated as "ease of operation" vs. "loss of energy" (due to drag), the Matrix offers the following inventive principles: *separation, periodic action* and "*heels over head*." These inventive principles refer to the following actions.

- *Separation*: single out a part or the system which interferes with the intended system's functioning; single out the useful function of the system.
- *Self-service*: make the system perform its own auxiliary functions; use wasted energy and/or materials available in the system.
- *Non-mechanical changes in systems:* replace a mechanical system with an optical or acoustical one; use electricity, magnetism and/or electromagnetism in the system; change uniform fields into non-uniform fields, permanent fields into variable fields, unstructured fields into structured fields; use magnetic and/or electromagnetic fields together with ferromagnetic particles.
- *Neutral environment:* use a neutral instead of a (re)active environment; use voids.
- "*Heels over head*": do the opposite of what seems to be required in the problem.
- *Periodic action*: use periodic or impulsive actions instead of continuous actions; use pauses in periodic actions to perform auxiliary actions.

It soon becomes apparent that *separation* is twice referenced, leading to the assumption that there is a high probability that the extrication of a part of the sealant could lead to a solution. In fact, it would be sufficient if the sealant just covered the gap between the panels. This would ease the process of closing the gaps, and would not degrade the aerodynamic properties of the aircraft skin.

The solution to the problem, indeed, reflects this approach: a metal cap with an elastomeric retainer snaps into the gap, thus making its installation very quick (Fig. A2.3).

Both the Matrix and the inventive principles are among the earliest tools of TRIZ developed by Altshuller. Although using these tools is straightforward, their effectiveness is relatively low. Example A2-1 clearly exposes the fundamental limitations of the problem solving approach based on these tools.

(1) Formulation of system conflicts is not structured: the Matrix itself offers no guidelines on how to correctly define system conflicts. As a result, many system conflicts can be suggested for the same initial situation; a correct system conflict, however, may not necessarily be among them.

(2) The finite set of predefined typical attributes in the Matrix often makes it difficult to associate a specific (unique) attribute of a particular system with some of the typical ones. In such a case, this specific attribute has to be "stretched," at times to beyond recognition, to fit the Matrix.

(3) For a given system conflict, the Matrix proposes certain inventive principles; it, does not, however, recommend any tactic for using these principles: which of the proposed inventive principles should be used? What part(s) of the system should these principles be applied to? Should these principles be applied separately or together? If the latter, is there a preferred order of their application (e.g., first, use principle X to modify the part, then apply principle Y)?

(4) The power of this approach is significantly lessened by not invoking the concept of physical contradiction. "This approach is not effective for solving complex problems . . . [Complex] problems should be analyzed in-depth, by revealing the physical cause of system conflicts" (Altshuller, 1984).

These limitations are completely overcome in the analytical approach of ARIZ. Despite their deficiencies, both the Matrix and the inventive principles are still used by some TRIZ practitioners, and they have been extensively presented in literature (e.g., Altshuller, 1997, 1999; Tate and Domb, 1997; Rantanen and Domb, 2002). They have also been incorporated in some TRIZ-based software products, e.g., Invention Machine's TechOptimizer™ and CREAX's Innovation Suite.

Appendix 3

Standard approaches to solving inventive problems

There are three classes of Standards.[1]
- *Standards on system modification* (construction and evolution of sufields).
- *Standards on system detection and measuring.*
- *Standards on application of the Standards.*

The Standards are organized in a set built according to the logic of evolution of sufields: *simple sufields → complex sufields → intensified sufields*. Sufields at any of these stages can migrate either to a higher-level system (i.e., they become *bi-* or *poly-sufields*) or to a micro-level (i.e., a whole system or its part is replaced with a pair "field–substance" capable of performing the required action).

---

[1] The following Standards are adapted from (Altshuller, 1988a).

A3 Algorithm for using the Standards

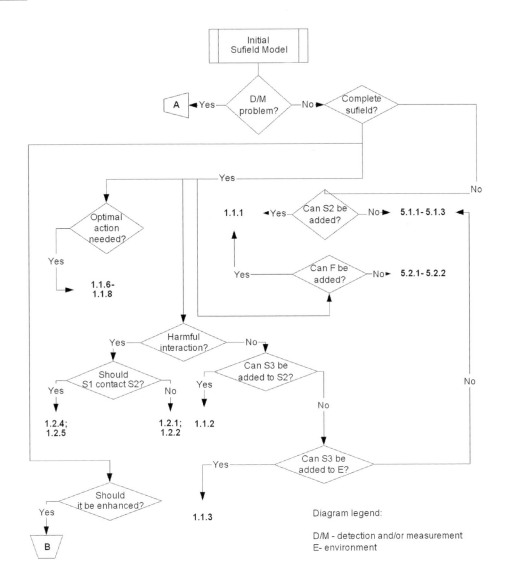

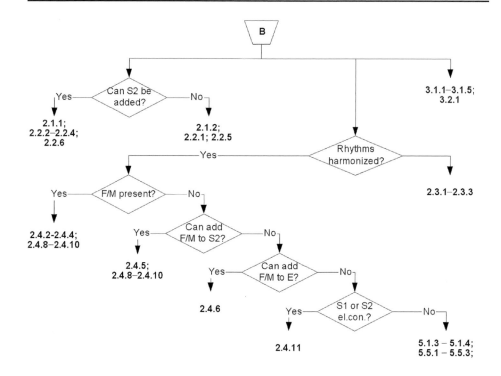

Diagram legend:

F/M - ferromagnetic material
el.con. - electroconductive

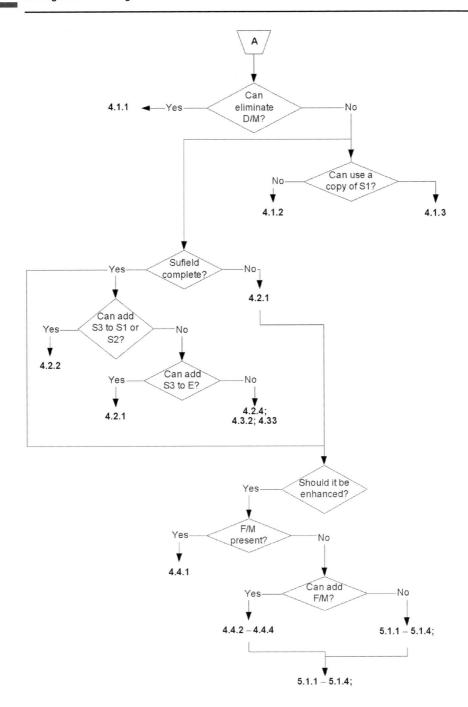

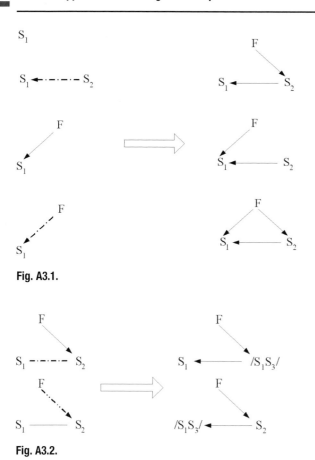

Fig. A3.1.

Fig. A3.2.

A3.1   Class 1: synthesis of sufields

A3.1.1   Group 1.1: improvement and introduction of useful actions

Standard 1.1.1

To enhance the effectiveness and controllability of an incomplete sufield, it must be completed by introducing the missing elements (Fig. A3.1).

Example: Problem 3.2.

Standard 1.1.2

If there is a sufield that is difficult to change as required, and there are no limitations on introduction of additives into the given substances, the problem can be solved by a permanent or temporary transition to an *internal compound sufield* (Fig. A3.2).

Here, S_1 – object; S_2 – tool; S_3 – additive; parenthesis indicate an internal compound connection (an external compound connection does not have brackets).

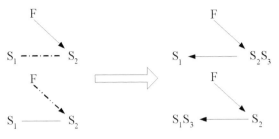

Fig. A3.3.

Fig. A3.4.

Example: To enhance heat- and mass-transfer with high-viscosity liquids, they are mixed with gas.

Standard 1.1.3

If there is a sufield that is difficult to change as required, and if there are limitations on adding additives into the given substances, the problem can be solved by a permanent or temporary transition to an *external compound sufield* (Fig. A3.3).

Example: To improve the weld quality, a welding rod is coated with an alloying material. As the coating burns, the weld is alloyed, and its strength increases.

Standard 1.1.4

If there is a sufield that is difficult to change as required, and there are limitations on the introduction or attachment of additional substances, the problem can be solved by using the present surrounding medium (environment) as the introduced substance (Fig. A3.4).

Example: Problem of self-unloading barge (p. 27).

Standard 1.1.5

If the surrounding medium does not contain a substance required to compose a sufield as in Standard 1.1.4, then this substance can be obtained by replacing this medium with another one, or by its chemical breakdown, or by introduction of additives.

Example: Poultry vaccination is a key part of the disease control program. Clearly, inoculating individual chicks would be a nightmare. The problem is solved by dispensing poultry vaccine into the drinking water.

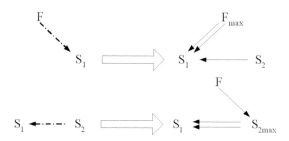

Fig. A3.5.

Fig. A3.6.

Standard 1.1.6

If a minimal (optimal) action is required that is difficult or impossible to carry out within the constraints of the problem, then the most intense action should be used, and its excesses should be removed (the excess of a field is removed by a substance, and the excess of a substance – by a field; Fig. A3.5).

Example: To apply a thin coat of paint on a part, it is dipped into a paint tank thus providing an excessive thickness of the coating. Then the part is spun, and the excess of paint is removed by centrifugal forces.

Standard 1.1.7

If there is a need to apply the maximal action to a substance, but it is unacceptable for some reason, the maximal action should be applied to another substance connected to the first substance (Fig. A3.6).

Example: Fabrication of pre-stressed reinforced concrete blocks requires stretching of steel reinforcing rods. This is done by heating the rods. After thermal expansion of the rods, they are fixed inside the concrete mass. However, if a wire is used instead of the more expensive rods, it has to be heated up to $700\,°C$, while only $400\,°C$ is tolerated before the wire starts losing its mechanical properties. It is suggested to heat an auxiliary heat-resistant rod which, after expansion, is attached to the wire. During the cooling process, the auxiliary rod shrinks and stretches the wire that remains cool.

Standard 1.1.8

If a selectively extreme action is required (e.g., an extreme action in certain areas while the minimal action is maintained in other areas), the field should have either maximal

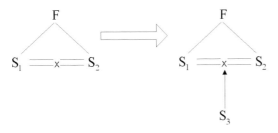

Fig. A3.7.

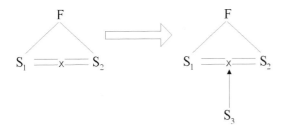

Fig. A3.8.

or minimal intensity. In the former case, the areas that have to be subjected to the minimal field are screened by a protective substance. In the latter case, the special substances generating a local field are introduced in the areas requiring maximal field; such substances could be, for example, termite mixtures for thermal action, explosive substances for mechanical action, etc.

Example: Problem 2.8.

A3.1.2 Group 1.2: breaking harmful actions

Standard 1.2.1

If both useful and harmful actions develop between two substances in a sufield, and it is not necessary to maintain a direct contact between the substances, the problem can be solved by introducing between the two substances a third substance from outside (this substance should be inexpensive or free; Fig. A3.7).

Example: Problem 3.3 (p. 60).

Standard 1.2.2

If both useful and harmful actions develop between two substances in a sufield, and it is not necessary to maintain the direct contact between the substances, but use of outside substances is prohibited or undesirable, the problem can be solved by introducing a third substance between the two, which is a modification of the two original substances (Fig. A3.8).

Example: Problem 2.10 (p. 41).

Fig. A3.9.

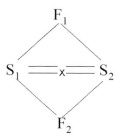

Fig. A3.10.

Standard 1.2.3

If there is a need to eliminate a harmful action of a field upon a substance, it can be achieved by introducing a second substance that draws off upon itself the harmful action of the field (Fig. A3.9).

Example: Problem 2.11 (p. 42).

Standard 1.2.4

If both useful and harmful actions develop between two substances in a sufield, and a direct contact between the substances should be maintained, the problem can be solved by transition to a double sufield, in which the existing field F_1 provides the useful action, while the harmful action is neutralized (or the harmful action is transformed into the second useful action) by a new field F_2 (Fig. A3.10).

Example: Problem 3.5 (p. 65).

Standard 1.2.5

If a sufield with a magnetic field has to be broken, the problem can be solved by using physical effects "turning off" ferromagnetic properties of substances by impact demagnetization or by heating above the Curie point (Fig. A3.11).

Example: Resistance welding of ferromagnetic powders to surfaces is often accompanied by repulsion of the powder by the magnetic field of the welding current. It was suggested to heat the powder up to its Curie temperature before bringing it into the welding zone.

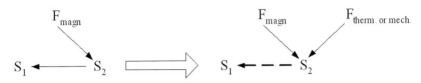

Fig. A3.11.

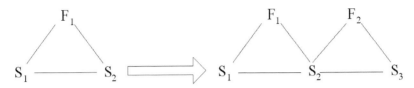

Fig. A3.12.

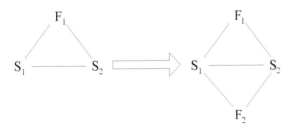

Fig. A3.13.

A3.2   Class 2: evolution of sufield systems

A3.2.1   Group 2.1: transition to complex sufields

Standard 2.1.1

If effectiveness of a sufield has to be enhanced, it can be achieved by transforming one substance of the sufield into an independently controlled sufield thus generating a *chain sufield* (Fig. A3.12).

Example: Example 5.44.

Standard 2.1.2

If controllability of a sufield has to be enhanced, but replacement of elements of this sufield is not allowed, the problem can be solved by synthesis of a *double sufield* by introducing a second well-controllable sufield (Fig. A3.13).

Example: Example 5.43.

A3.2.2   Group 2.2: intensification of sufields

Standard 2.2.1

The effectiveness of a sufield can be enhanced by replacement of an uncontrollable or poorly controllable field with a well-controllable field.

Examples: Replacement of gravitational field with mechanical, of mechanical with electric, etc.

Standard 2.2.2

The effectiveness of a sufield can be enhanced by increasing the degree of fragmentation of a substance acting as a tool.

Example: Example 5.16.

Standard 2.2.3

A special case of fragmentation of a substance is transition from solid substances to capillaries or porous substances.

This transition progresses along the line of evolution: *solid substance → solid substance with one cavity → solid substance with multiple cavities (perforated substance) → porous or capillary filled substance → porous or capillary filled substance with orderly structured (and/or calibrated) capillaries or pores.*

Example: In soldering electronic parts on a circuit board, a soldering defect such as a solder bridge (excess solder), which causes terminals to be short-circuited, sometimes occurs and needs to be removed. US Patent 6,681,980 describes a soldering iron with a tip that has capillary-like grooves that suck up the excess molten solder.

Standard 2.2.4

The effectiveness of a sufield can be enhanced by increasing its flexibility, which means transition to a system structure that is more flexible and capable of fast changes.

Examples: See the examples illustrating the law of increasing dynamism in Chapter 5.

Standard 2.2.5

The effectiveness of a sufield can be enhanced by transition from uniform or disordered fields to non-uniform fields or fields with a particular spatial–temporal structure (stable or variable; Fig. A3.14).

Example: US Patent 5,006,266 describes using ultrasonic standing waves to separate particulate materials in liquids by attracting the individual particles to the nodes or the antinodes.

Fig. A3.14.

Fig. A3.15.

Standard 2.2.6

The effectiveness of a sufield can be enhanced by transition from homogeneous substances to substances with a particular spatial-structured structure (the latter may be stable or variable; Fig. A3.15).

Example: To produce porous flame-resistant blocks, the directional pores are generated by burning silk threads placed into the green mass.

A3.2.3 Group 2.3: harmonization of rhythms

Standard 2.3.1

The effectiveness of a complete sufield can be enhanced by tuning or detuning the frequency of a field action with the natural frequencies of the object or the tool.

Example: To extract stones from the urethra, sometimes a string with a loop is introduced into the urethra. The stone is pulled out with the loop. It is suggested to apply a pulsating force with the frequency being a multiple of the frequency of perilstatic movements of the urethra. This enables a reduction in the trauma of the procedure, by reducing pain, and assists removal of stones of various shapes and sizes.

Standard 2.3.2

The effectiveness of a complex sufield can be enhanced by tuning or detuning the frequencies of fields in relation to one another.

Example: There is a method of enriching fine-milled magnetic ores in which the ore is acted upon by a combination of a traveling magnetic field and vibration. The effectiveness of the separation process is enhanced by varying the intensity of the magnetic field synchronously with the vibration.

Fig. A3.16.

Standard 2.3.3

If two actions, such as change and measurement, are incompatible, then one action is performed during pauses of the other action (generally, pauses in one action should be filled by another useful action).

Example: To provide for automatic control of the thermal cycle of resistance spot welding, it is necessary to measure the thermoelectric potential. The control accuracy for high pulsating frequency of welding is improved by measuring the thermoelectric voltage potential during pauses between pulses of the welding current.

A3.2.4   Group 2.4: fefields and efields

The highest degree of intensification in modern technological systems can be applied to *fefields* (sufields with dispersed ferromagnetic substances and magnetic field.

Standard 2.4.1

The effectiveness of a sufield system can be enhanced by using a ferromagnetic substance and magnetic field (Fig. A3.16).
   This Standard is for application of a ferromagnetic substance that is not finely fragmented. These systems can be called *proto-fefields* – structures in transition towards fefields. The Standard is applicable to simple as well as to complex sufields.

Example: A hammer equipped with a magnetic insert can pick up nails and tacks and hold them prior to their use.

Standard 2.4.2

The controllability of a sufield or a proto-fefield can be enhanced by replacing one of the substances with ferromagnetic particles or by adding ferromagnetic particles, such as chips, granules, grains, etc. and by using magnetic or electromagnetic fields.
   After becoming a fefield, the sufield system repeats the evolution cycle of sufields, but at a new, higher level, since fefields are characterized by high controllability and effectiveness.

Example: Magnetic tapes for therapeutic application are typically plastic foils with magnetic particles embedded in the plastic. These magnetic particles are aligned by

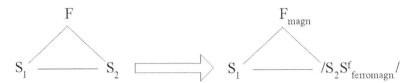

Fig. A3.17.

applying an external magnetic field to the foil; this produces a magnetically polarized area.

Standard 2.4.3

The effectiveness of a fefield can be enhanced by transition to a magnetorheological fluid.

Example: Example 5.13 (p. 126).

Standard 2.4.4

The effectiveness of a sufield can be enhanced by using a capillary or porous structure.

Example: Impregnating powder metal parts, which have porous surfaces, with magnetorheological fluids enhances the binding of the parts together in various mechanisms (US Patent 6,428,860).

Standard 2.4.5

If there is a need to enhance controllability by transition to fefield, but replacement of the substance by ferromagnetic particles is not acceptable, then the transition is performed by constructing an internal or external complex fefield by adding ferromagnetic particles to one of the substances (Fig. A3.17).

Example: In magnetic–abrasive machining, common abrasive particles sintered with iron grits are placed between an electromagnet and the workpiece. When there is a relative motion between the particles and the workpiece, the abrasion takes place (Jain, 2002).

Standard 2.4.6

If controllability should be enhanced by transition from a sufield to a fefield, but replacement of the substance by ferromagnetic particles or additions to the substances is not acceptable, then ferromagnetic particles should be added to the surrounding medium (environment).

Example: A plastic sheet impregnated with magnetized particles ("magnetic mat") is used in a variety of environments (workshops, garages, etc.) to keep tools, nuts, bolts, and other parts within easy reach.

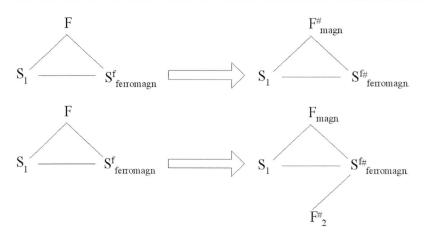

Fig. A3.18.

Standard 2.4.7

The controllability of a fefield system can be enhanced by using physical effects.

Example: To enhance the sensitivity of measuring magnetic amplifiers, the absolute temperature of their core is maintained at 0.92–0.99 of the Curie temperature of the core material.[2]

Standard 2.4.8

The effectiveness of a fefield system can be enhanced by its dynamization, or transition to a flexible, variable structure of the system.

Example: The thickness of the wall of a hollow object is measured using a magnetic transducer outside and a ferromagnet inside of the object. To increase measurement accuracy, the ferromagnet is given the shape of an inflatable elastic balloon coated with ferromagnetic particles, so it can exactly conform to the object being measured.

Standard 2.4.9

The effectiveness of a fefield system can be enhanced by transition from fields having a homogenous or non-orderly structure to fields having heterogeneous or defined 3-D structure (constant or variable; Fig. A3.18). In particular, if a substance within a fefield (or a substance which may become a part of a fefield) has to be given a certain three-dimensional structure, then the process should be performed in the field whose structure corresponds to the required structure of the substance.

Example: To fabricate bristled plastic mats, ferromagnetic powder is added to the molten plastic according to the required bristle pattern. A magnetic field pulls out the powder thus forming the bristles. As an additional benefit, this solution allows for fast development and alteration of the bristle patterns.

---

[2] Before reaching the Curie temperature, the magnetic susceptibility of ferrites reaches its maximum value (Hopkinson effect).

Standard 2.4.10

The effectiveness of a "proto-fefield" or fefield system can be enhanced by harmonization of rhythms of constitutive components of the system.

Examples: A "toss-up" vibratory regime is used to convey bulk and chunk ferromagnetic materials. The conveyance rate is increased by applying a pulsating magnetic field that travels in the direction of conveyance. The duration of the magnetic pulses is adjusted to be equal to the duration of the "flying phase" of the material being vibrated.

Standard 2.4.11

In case of difficulties with introduction of ferromagnetics or with providing magnetization, interaction of an external electromagnetic field with directly conducted or induced currents, or interaction between these currents should be used.

Example: Gripping metallic non-magnetic parts is done by passing a direct electric current through the part body in the direction perpendicular to the field strain lines of a magnet which is a part of the gripping device.

Fefields are systems into which ferromagnetic particles are introduced, while *efields* are systems in which electric currents act (or interact) instead of ferromagnetic particles. Evolution of efields, similarly to that of fefields follows the common line: *simple efiels → complex efields → efields on an external medium → dynamization → structuring → harmonization of rhythms.*

Standard 2.4.12

If a magnetic field cannot be used, one can use an electrorheological fluid.

Examples: See Chapter 5, section entitled *Smart materials*.

A3.3 Class 3: transition to a higher-level system and to micro-level

A3.3.1 Group 3.1: transition to bi- and poly-systems

Standard 3.1.1

The effectiveness of a system at any stage of its evolution can be enhanced by combining the system with another system(s) into bi- or poly-system.

Examples: reading glasses, two-barrel hunting rifle, semiconductor diode, candy with liquor, multi-ink pen, roll of postal stamps, multi-layer cake, etc.

Standard 3.1.2

The enhancement of effectiveness of synthesized bi-systems and poly-systems is achieved, first of all, by evolution of connections between components of these systems.

Example: Folding reading glasses.

Standard 3.1.3

The effectiveness of bi- and poly-systems is enhanced when diversity of their components is increasing along the line: *identical components → components with shifted characteristics → different components → inverse combinations 'component and anti-component.*

Examples: Examples 5.22–5.26 (pp. 136–138).

Standard 3.1.4

The effectiveness of bi-systems and poly-systems is enhanced in the process of their convolution. The main reason for this is the reduction of auxiliary parts, e.g., a double-barreled gun has only one butt. Completely convoluted bi- and poly-systems again become mono-systems and the evolutionary cycle may repeat itself on the new level.

Examples: Examples 5.27–5.30.

Standard 3.1.5

The effectiveness of bi- and poly-systems can be enhanced by distributing incompatible (contradictory) properties between the system and its components: use a two-level system in which the whole system has a property P while its components have an opposite property −P.

Example: Problem 2.7 (p. 37).

A3.3.2   Group 3.2: transition to micro-level

Standard 3.2.1

The effectiveness of a system at any stage of its evolution can be enhanced by the transition from macro-level to micro-level, when the system or its part is replaced by a substance capable to perform the desired action when controlled by a field.

Example: Example 5.37.

A3.4   Class 4: standards for detecting and measurement

A3.4.1   Group 4.1: Detours

Standard 4.1.1

If there is a problem requiring measurement or detection in a system, then the system should be changed so that the need to measure or detect is eliminated.

Example: Problem 3.6 (p. 67).

Standard 4.1.2

If Standard 2.1.1 cannot be used, then it is worthwhile to replace direct interaction with the object by interaction with its copy or photographic image.

Example: Multi-spectral satellite imagery is an effective tool for crop assessment. It is based on the ability of plants to reflect light in the near-infrared part of the spectrum. Plants' infrared reflectance increases with the growth of the vegetative biomass. Satellites measure this reflectance and convert it to the digital data that are correlated to the amount of vegetative biomass and potential crop yields.

Standard 4.1.3

If Standards 4.1.1 and 4.1.2 cannot be used, then it is advisable to transform the problem into one requiring sequential detection of changes.

Any measurement has a certain degree of accuracy. Thus, in measurement problems even if a continuous measurement is required, it is always possible to separate an elementary measurement act into two consequent detections.

Examples: In the "chamber method" for mining copper ore, huge underground halls or "chambers" are generated. Due to blasts and to other reasons, the ceiling of the chamber may peel off in some areas and fall. It is necessary to monitor the condition of the ceiling and to measure the developing "pits". However, it is very difficult since the ceiling is at the height of a five-storey building. It is suggested to drill horizontal holes above the ceiling in the preparation process for the construction of the chambers, and to insert into the holes luminescent substances of various colors. If in some location the ceiling develops a cupola-like pit, it can be easily detected by the exposing luminofore. The depth of the pit can be assessed by the luminofore's color.

A3.4.2 Group 4.2: synthesis of measurement sufields

Standard 4.2.1

If it is difficult to measure or detect elements of an incomplete sufield, and Standard 4.2.1 cannot be used, then this sufield has to be completed and have a field as its output (Fig. A3.19).

Example: Problem 3.7 (p. 69).

Standard 4.2.2

If a system or its part is not amenable to a detection or measurement procedure, then the problem can be addressed by transition to an internal or external complex sufield by introducing additives which are easier to detect (Fig. A3.20).

Example: The actual area of contact between two parts is measured by painting one of the contacting surfaces with a luminescent die.

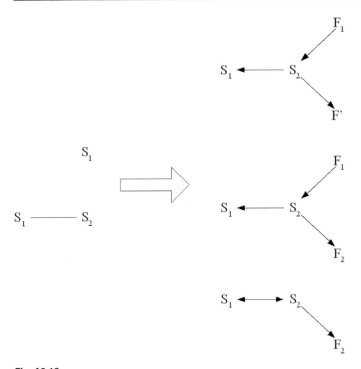

Fig. A3.19.

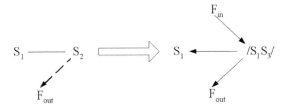

Fig. A3.20.

Standard 4.2.3

If a detection or a measurement procedure is difficult to perform at a certain moment of time and there is no possibility to introduce an additive generating an easily detectable and/or measurable field, then the additive has to be introduced into the external medium. By monitoring the latter, it might be possible to record changes in the state of the object of interest (Fig. A3.21).

Example: To monitor wear in an internal combustion engine, it is required to measure the amount of "worn-out" metal. Its particles are suspended in the lubricating oil. It is suggested to add a luminophore to the oil: the metal particles would suppress the glow.

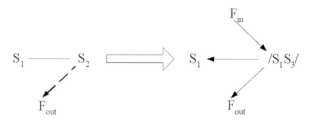

Fig. A3.21.

Standard 4.2.4

If an additive cannot be added to the external medium, as by Standard 4.2.3, then these additives might be generated in the sufield's substance, e.g. by its dissociation into constitutive elements or by changing its aggregate state. As such an additive, gas or steam bubbles obtained by electrolysis, cavitation, etc., are frequently used.

Example: Cavitation is used to measure a liquid's flow velocity in a pipe (if introduction of additives into the fluid is prohibited). Small (and thus stable) bubbles are detected either with the help of the inductance or capacitance sensors.

A3.4.3 Group 4.3: intensification of measurement sufields

Standard 4.3.1

The effectiveness of detection and measurement within a sufield system can be enhanced by using physical effects.

In some cases it is desirable that substances constituting a sufield make a thermo-couple which provides "free" information about the state of the system's parameters. An "information field" can also be generated by induction.

Example: Problem 3.13 (p. 76).

Standard 4.3.2

If certain changes in a system cannot be detected or measured directly, and a field cannot be applied to the system, then the problem may be solved by excitation of resonance vibration in the system or its part; the changes in the system can be monitored by recording the changes in the vibration frequency.

Example: The mass of oil in a large stationary oil reservoir can be measured by imparting forced vibrations to the reservoir.

Standard 4.3.3

If Standard 4.3.2 cannot be applied, then condition of the system may be assessed by monitoring changes in the natural frequency of an object (external medium), connected to the system to be controlled.

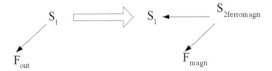

Fig. A3.22.

Example: The mass of a clinker in a cement baking kiln is determined by monitoring variation of amplitude of self-excited vibrations of the gas over the pseudo-boiling layer.

A3.4.4   Group 4.4: transition to measurement fefields

Standard 4.4.1

Measurement sufields with non-magnetic fields have a tendency of transition into proto-fefields (Fig. A3.22).

Example: US Patent 6,742,229 describes a buckle device used in automotive seat belt systems, in which it can be detected whether or not the "tongue" inserted into a holder is locked. This is accomplished with the help of a magnetic plate, which moves together with the "tongue" and a magnetic sensor that detects the plate's position.

Standard 4.4.2

To enhance the effectiveness of detection or measurement by proto-fefields and sufields systems, it is desirable to introduce fefields by replacing one of the substances with ferromagnetic particles (or by adding ferromagnetic particles) and then performing detection or measurement of the magnetic field.

Example: The degree of hardening (solidification) or softening of polymeric compositions can be non-destructively evaluated by adding magnetic powder into the composition. The measured parameter is the change of the composition's magnetic permeability in the process of solidification/softening.

Standard 4.4.3

If Standard 4.4.2 cannot be used, then transition to a fefield may be executed by constructing a complex fefield by introducing additives to the substances.

Example: Hydrocrushing of rock in mining is performed by applying a pressurized liquid to the rock. The fluid's parameters are monitored by adding ferromagnetic powder to it.

Standard 4.4.4

If Standard 4.4.3 cannot be used, then ferromagnetic particles have to be introduced into the external medium.

Example: To study waves generated by a model ship, magnetic particles are added to the water.

Standard 4.4.5

To enhance the effectiveness of a fefield measurement system, one can use physical effects, such as transition through the Curie point, Hopkinson effect, Barkhausen effect,[3] etc.

Example: US Patent 4,534,405 discloses a method for inspecting the surface of steel stock. A thin surface layer of the steel stock is intensively cooled on the surface to be inspected, to a temperature below the Curie point. The core of the steel remains hot. Immediately thereafter, a magnetic field is induced in the cooled surface layer and disturbances in the induced field caused by defects are detected.

A3.4.5 Group 4.5: directions of evolution of measurement sufields

The evolution of measurement sufields progresses along conventional system transitions but it is also characterized by some specific features.

Standard 4.5.1

The effectiveness of a measurement system at every phase of its evolution can be enhanced by transition to a bi- or poly-system.

Example: To measure the body temperature of small insects, a multitude of them are placed in a glass. Then, a conventional thermometer can be used.

Standard 4.5.2

Measurement systems evolve in the direction: *measuring of a function* → *measuring the first derivative of the function* → *measuring the second derivative of the function*.

Example: Vibration measurement techniques have evolved from measuring the amplitude of vibratory displacement (seismometer) to measuring vibratory velocity (velocimeter) to measuring vibratory acceleration (accelerometer). At present, the overwhelming majority of vibration measurements are performed using accelerometers.

---

[3] This is an abrupt increase of the value of the magnetic field during magnetization of a ferromagnetic material.

A3.5   Class 5: standards on application of the standards

A3.5.1   Group 5.1: introduction of substances

Standard 5.1.1

If there is a need to introduce a substance into the system, but it is unacceptable due to the problem specifications or to performance conditions of the system, then some indirect ways should be used.

(1) *Use "voids" instead of substances.*

Example: Problem 3.8 (p. 71).

(2) *Introduce a field instead of a substance.*

Example: Problem 3.11 (p. 73).

(3) *Use an external instead of an internal additive.*

Example: To measure the wall thickness of a hollow ceramic vessel it is filled with a highly electroconductive liquid, one lead of the ohmmeter is dipped into the liquid, and another lead touches the external surface at the point where thickness is to be measured.

(4) *Introduce a micro-dose of an extremely active additive.*

Example: A mineral oil-based lubricant is used for drawing pipes. To reduce hydrodynamic pressure of the lubricant in the deformation zone, 0.2–0.8% by weight of polymethilacrylate is added to the lubricant.

(5) *Introduce a microdose of a conventional additive but strategically concentrate it in critical segments of the object.*

Example: To make an electroconductive polymer, ferromagnetic particles, shaped as discrete fibers, are introduced into the polymer and oriented in the desired direction of high conductivity.

(6) *Introduce an additive for a short duration.*

Example: Orientation of hollow ferromagnetic components is enhanced by inserting ferromagnetic pieces into the components for the duration of the orientation procedure.

(7) *Use, instead of the object, its substitute (mock-up) into which insertion of an additive is permissible.*

Example: To enhance accuracy of three-dimensional analysis of the cross-sectional shapes, the latter are simulated by the horizontal surface of a liquid inside a transparent mock up which can be given various positions and inclinations.

(8) *To introduce an additive as a component of a chemical composition from which the additive is subsequently extracted.*

Example: Plastification of frictional wooden surfaces during their sliding is achieved by impregnating the wooden parts with a chemical composition that dissociates at the friction temperature and emits ammonia.

(9) *To generate an additive by dissociating the surrounding medium or the object itself,*
   e.g., by electrolysis or by changing the aggregate state of a part of the object or of
   the surrounding medium.

   Example: To intensify removal of a residue generated by the electrodischarge
   machining process, the gas is generated by electrolysis of the electrolyte next to
   the machining zone.

Standard 5.1.2

If a system is responds poorly to the desirable changes, but the tool cannot be replaced
or additives introduced, then the tool is replaced by the article. The latter is divided into
parts interacting between themselves.

Example: Electrostatic coagulation of dust in mines is enhanced by splitting the dust
stream into two parts each of which is oppositely charged, and directing them toward
each other.

Standard 5.1.3

A substance introduced into the system, after it has performed its function, should
disappear or become indistinguishable from the substance that had already been present
in the system or in its environment.

Example: Problem 3.9 (p. 72).

Standard 5.1.4

If there is a need to introduce a large quantity of a substance, but it is prohibited by
the specifications of the problem or is unacceptable by performance conditions, then a
void (inflated structures or foam) is used as the substance.

Example: Using inflatable structures as temporary buildings.

A3.5.2 Group 5.2: introduction of fields

Standard 5.2.1

If there is a need to introduce a field into a sufield system, then the first of the already
present fields have to be used. Substances comprising the system are sources of these
fields.

Example: Problem 3.12 (p. 75).

Standard 5.2.2

If there is a need to introduce a field into a sufield system, and Standard 5.2.1 is
inapplicable, then fields available in the environment should be used.

Example: In 1999, Citizen Watch Company introduced watches powered by the temperature difference between the back of the watch's case, warmed by body heat, and the surrounding air.

Standard 5.2.3

If a field needs to be introduced into the system, but Standard 5.2.1 cannot be used, then the field should be used whose carriers or sources can be substances already present in the system or in the environment.

Example: The system "machined part-cutting tool" is used as a thermocouple for measuring cutting zone temperature ("natural thermocouple").

A3.5.3   Group 5.3: phase transitions

Standard 5.3.1

The effectiveness of the use of a substance can be enhanced, without introduction of other substances, by *phase transition 1*, i.e. by changing the phase state of the present substance.

Example: Power for pneumatic systems in mines is supplied by liquefied, rather than compressed, gas.

Standard 5.3.2

"Dual" properties can be realized by use of *phase transition 2*, i.e., by employing substances capable of transition from one phase to another depending on the work conditions.

Example: Use of shape memory alloys (e.g., Example 5.17) (p. 130).

Standard 5.3.3

A system's effectiveness can be enhanced by use of *phase transition 3*, i.e. phenomena accompanying a phase transition.

Example: A handling system for frozen goods has supporting elements made of ice blocks (friction is reduced due to the ice melting).

Standard 5.3.4

"Dual" properties of a system can be realized by *phase transition 4* (i.e., replacing a single-phase state by a two-phase state).

Example: Cleaning filters for delicate products, such as caviar, involves breaking lumps of the product and then washing out the contaminants with water. To reduce the potential damage to the product, the water is saturated with gas.

Standard 5.3.5

The effectiveness of technological systems resulting from *phase transition 4* can be enhanced by introduction of physical or chemical interaction between parts or phases of the systems.

Example: To enhance gas compression in the cooler and expansion in the heater of a two-phase working medium for compressors and thermal machinery consisting of gas and fine solid particles, sorbents with general or selective sorption ability are used as materials for the solid particles.

A3.5.4 Group 5.4: specifics of using physical effects

Standard 5.4.1

If a substance should periodically exist in different physical conditions, the transition should be performed by the object (substance) itself by using reversible physical transformation such as phase transitions, ionization/recombination, dissociation/association, etc.

Example: Problem 2.5 (p. 35).

Standard 5.4.2

If a strong output action is needed in response to a weak input, then the transducer material should be brought to a near-critical condition. Then even a weak signal becomes stronger.

Example: To test a product, it is immersed in a liquid without gaseous inclusions. Then a pressure differential is generated between the product's cavity and the environment above the liquid level. The hermeticity is monitored by detecting gas bubbles in the liquid. To enhance the sensitivity of the test, the liquid during the testing process is maintained in an overheated condition.

A3.5.5 Group 5.5: obtaining particles of substances

Standard 5.5.1

If particles, such as ions, are needed to solve the problem but their direct generation is impossible due to the problem's specifications, the required particles should be obtained by decomposing a substance of a higher structural level, such as molecules.

Example: Problem 2.5 (p. 35).

Standard 5.5.2

If certain particles of a substance (e.g., molecules) are needed that are impossible to obtain directly or by using Standard 5.5.1, then these particles should be generated by building up or combining particles of a lower level, such as ions.

Example: Nanotechnology – assembling products from individual atoms (Chapter 5, section *Increasing Non-Uniformity of Materials*).

Standard 5.5.3

The simplest way of applying Standard 5.5.1 is to destroy the nearest higher "whole" or "excess" (negative ions) level, and the simplest way of applying Standard 5.5.2 is to complete build-up of the nearest lower "incomplete" level.

Example: Chemical vapor deposition (CVD) is widely used in the semiconductor industry to deposit various films on substrates. In CVD processes, the coating is deposited from a reactive gas. The chemicals (molecules) decompose (into atoms) at the high temperature in the reactor and recombine to form the desired coating (again, molecules) on the hot substrates.

Appendix 4

Using TRIZ in management practice

TRIZ was originally developed as a tool for enhancing engineering creativity. As TRIZ evolved, it became clear to Genrikh Altshuller and his followers that its fundamental principles should be applicable to other human pursuits such as science, art and others (Altshuller, 1963, 1988b; Filkovsky, 1974). The idea was to apply the same approach used while developing TRIZ (i.e., treating all entities of study as evolving systems, accumulating a large database of "strong solutions", formulating laws of evolution specific to that given class of entities, etc.) to analyze other evolving systems. Natural objects for such analysis would be the ones most closely related to technology – managerial and business systems. The following is an outline of TRIZ concepts and tools that might be used in business and organizational contexts.

A4.1 Tools of TRIZ and management issues

There are many occurrences of the spontaneous use of the laws of evolution by businesses. For example, the law of increasing dynamism is manifested in such approaches as outsourcing, the postponing of product differentiation (delegation of final product assembly to dealers and distributors to reduce the delivery time with customer-required modifications and options [Davis and Sasser, 1995]), the use of flexible production systems (McGraw, 1996), the use of reconfigurable machining systems and modular tooling structures, and others.

Another business application of increasing dynamism is eloquently described: "Small jets are starting to do to the airline industry what PCs did to mainframe computing; fractional horsepower machines did to turbines; minimills did to steel; cellular is doing to telephony; mutual funds are starting to do to centrally managed corporate and government pension plans, and eventually will do to Social Security, and what coming; mini-generators will do to massive power plants – give customers more service, more flexibility, more control at less cost, as well as generate new products and services.

It's about power moving away from the machine-age center toward individuals of the microchip era". (Forbes, 1997)

The same law of evolution can be identified with the ongoing societal trend of delegation of more and more authority to lower levels of government/management (from federal to state governments, from top levels to lower levels of management, etc.). This trend is also predicted by the law of shortening energy flow path.

A good illustration of the law of transition to higher-level systems is Bloomberg Information Television, which features a multi-screen presentation: the anchor reading the principal news occupies the largest square on the screen, two still and constantly upgraded squares provide secondary information such as horoscopes, weather, stock exchange performance, etc., all while one continuously moving rectangle provides stock quotes, sport scores, etc. Modular systems, both in products and in manufacturing systems, also represent the transition to poly-systems (as well as increasing dynamism, as described above).

The developments in these examples took place without the deliberate use of TRIZ; they are just the objective trends predicted by TRIZ. One can imagine how much more could be achieved if the laws of evolution were directly and explicitly applied to various strategic management problems.

Problems and bottlenecks associated with managing large business systems are usually solved solely by management approaches: adding personnel, offering incentives, acquiring more, or more advanced, equipment, etc. There are many instances when TRIZ approaches can be used, however. One example of solving a management problem in the TRIZ spirit is the oversight of toll collection for bridges/tunnels. In the past, there were toll booths in both directions for bridges and tunnels leading to/from Manhattan (New York City). Once it was understood that vehicles coming to the island would, in the overwhelming majority of cases, also leave it, the function of collecting tolls from outbound vehicles was assigned to the inbound toll booths. This was done by increasing the amount of the inbound toll, and closing the outbound booths. This management action replaced the real system (outbound booth with a toll collector) with an ideal system – no physical entity, but performing the required function. This action can also be presented as a realization of the law of transition to a higher-level system (convolution of a bi-system into a higher level mono-system).

There are numerous other instances when complex management problems could be solved in a cost-effective way by creatively resolving an engineering problem that creates the bottleneck. The most effective approach is by addressing the mini-problem, where the solution is achieved with minimal changes to the infrastructure. A good example is the problem of transporting hot molten slag from a blast furnace to a slag granulating plant (Altshuller, 1988b).

This was initially perceived as a purely management problem. Initially, the slag was conveyed in open ladles, mounted on railway platforms. On the way, the slag cooled down, and a thick, hard crust developed on its surface. To unload the liquid slag, it was necessary to punch holes through the crust. Also, large solidified residues of slag in the ladles had to be manually removed before reloading the ladle at the blast furnace. In both these operations, low-skilled workers were exposed to elements, to dust, and to very hard labor. Nobody volunteered for these jobs, and financial incentives did not help. Another approach considered was to build two expensive crane stations at the loading and unloading sites and to fabricate many expensive and very heavy lids for the 10–20 ft diameter ladles. A TRIZ specialist invited to a high-level management meeting proposed an engineering solution. The solution (sprinkling the slag surface with water, thus generating an effective lid by foaming the top layer of the molten slag) completely solved the complex management problem, with minimum expenditures.

As this book was being written, a student taking the TRIZ course at Wayne State University brought a problem for the class to solve. The student works for an automotive supplier that develops and manufactures the brake caliper assembly, a sub-assembly of the car brake. The primary function of the caliper assembly is to direct the brake fluid from the master cylinder to the caliper. The current assembly consists of four separate parts. The operators assembling the parts often accidentally leave one of them out. This causes the brake fluid to leak, which is clearly undesirable.

As in the previous case study, conventional managerial tricks were to no avail. A simple TRIZ analysis of the situation showed that the part that was repeatedly left out was an auxiliary correcting tool. Eliminating that part and assigning its function to one of the remaining parts was suggested. The proposed design was approved by the company's management, and a detailed development is now under way.

Although not every management problem can be solved this way, the potential gains from at least attempting such solutions are enormous. Even the 20% success rate is worthwhile.

Some tools of TRIZ can clearly be applicable to socio-business interactions (e.g., those between managers and their subordinates). As of now, there are not too many examples illustrating this statement. One example is the application of the sufield analysis to training. In a complete sufield, the field acts on the object not directly, but through another substance (tool).

While training and disciplining members of a working team, it is often not very productive to interact directly with the "object employee." Frequently, it is better to say unpleasant observations (or advice) to a third person, while the "object employee" is present. This "information field" from a manager or an instructor acts upon the third person, who acts as a "tool." With such an approach, the "object" usually "gets the message" more readily, without developing a negative attitude. A similar approach is sometimes practiced by medical doctors for communicating with "difficult" patients (Lown, 1996).

A4.2   Applications

The authors were involved in several specific projects on the application of TRIZ to organizational/managerial issues. Below are brief accounts of some of these projects.

Case story 1

A global corporation, headquartered in a European country, owns dozens of manufacturing companies spread across several continents. The corporation has a large group of in-house management consultants who are involved in various projects with the constituent companies. The group is based at the corporate headquarters and its consultants are routinely on the road, traveling far to work on their projects, which may last anywhere from a week to a couple of years. As a rule, the consultants work with different companies – having completed a project, say, with a Brazilian company, the consultant may be assigned to a project in Italy or Canada, or in any other country.

While working on various assignments and in diverse business and cultural environments, the consultants acquire extensive experience, which may benefit many of their colleagues. But here lies the problem. To be proficient in local business and cultural traditions, a consultant must spend much time in that particular country (or with the individual company). Ideally, he or she should be a native resident, fluent in the local language and customs. When working in the specific conditions, however, one can only obtain limited experience. Yet today's ever changing business and technology environments constantly produce non-typical and urgent challenges. While indispensable, such experience may not offer adequate and timely (sometimes, in a matter of hours) solutions. To expand one's horizons and master new skills, one has to travel to many places and spend significant time and efforts within different environments. This approach, however, is difficult to carry out practically. Moreover, in this case, one does not have time to develop a sound understanding of new technologies and local business, and cultural traditions.

In order to share their experiences, the consultants gather at annual conferences, at which they report on the projects they have overseen, and talk about the tricks of the trade, thus enhancing each other's skills. Although effective, these conferences cannot fulfill the consultants' growing needs for the expeditious exchange of information and ideas.

TRIZ analysis

Let us start the analysis with the formulation of a system conflict: "if a consultant constantly stays in a particular country (or with the same company), then he/she obtains important specific knowledge, but limits his/her experience. If the consultant participates in various projects in different countries (companies), then he/she acquires diverse experiences, but never obtains profound specific knowledge".

In order to reveal the core of system conflicts, TRIZ recommends intensifying them by driving the competing requirements to their logical extreme. In this case, the intensification of the conflict is: *the consultant should permanently stay in the same country (with the same company) and, on the other hand, constantly move from one country (company) to another.* This sounds like a physical contradiction that can be overcome by using the separation principles. A simple analysis of this contradiction shows that it can be resolved by separating the contradictory demands in "space": the consultant "partly" stays within the same environment (company and/or country) and is "partly" present everywhere else. Of course, nobody is going to physically divide the consultant into parts. The solution will become more palatable if we recall the Ideality Principle, and employ it to assist us in resolving the contradiction. What we need is not the consultant as a physical entity, but his or her knowledge, expertise, experience (i.e., the function). This "part" of the consultant can be "separated" from its physical carrier, and easily transferred to all interested parties.

Solution
Using the available corporate IT resources to develop an on-line forum for the exchange of information (and, consequently, a database of projects and solutions) was suggested. This forum is used for posting questions, inquiries, messages, ideas and solutions. It allows for the fastest and the most comprehensive internal consulting on various problems: a consultant who posts a question/problem at the forum immediately becomes inundated with ideas and suggestions coming, literally, from all corners of the world. Now, a consultant can stay in the country (the company) as long as it takes to become an expert in the local culture, while having an opportunity for the instant acquisition/transfer of knowledge from/to other consultants.[1]

Case story 2
Two financial institutions in a European country – a large insurance company and a major bank – joined resources to form a larger entity. This merger resulted in the increased profitability and competitiveness of the new company, but also begat an unexpected problem.

After the merger, the new company slashed costs by laying off quite a few employees. This caused a redistribution (integration) of functions: a former actuary also became an investment authority and a former broker began to consult in actuarial matters. This arrangement has not affected the relations of the new company with its corporate clients, but it has diminished the loyalty of individual customers. The latter, in that country, have an age-old tradition of consulting on their investment needs with banks, and on their actuarial problems with insurance agents. A broker–actuary was perceived as neither an expert broker nor a professional actuary. As a result, the number of individual clients rapidly began to drop – they turned to the competitors who still respected the custom of strict specialization.

TRIZ analysis
This is a typical situation developing when systems undergo transitions "prescribed" by the laws of evolution. The merger of two business entities is a step along the line

---

[1] This solution – an Internet forum – may look conventional, but in 1995, when it was proposed, it was quite innovative.

of evolution *mono-bi-poly*. While evolutionary advantageous, this step is frequently associated with the emergence of system conflicts caused by an integration of different mono-systems having diverse natures (functions). The new bi-system cannot adequately function and evolve until these system conflicts are resolved.

The system conflict in our case is: "to reduce operational costs, company associates should combine the duties of actuaries and brokers, but this alienates the individual clients. If investment and actuarial consulting are assigned to different specialists, this could restore the loyalty of the individual clients, but would drive the costs up".

By analogy with the previous case, we can intensify the system conflict and formulate a physical contradiction: *an associate must be both a full-time broker and a full-time actuary*. Obviously, this contradiction cannot be overcome by separating the opposing demands in time, so it should be resolved by separating them in space (parts).

Solution

In any professional discipline, a body of knowledge is composed of two equally important parts: typical cases and situations (i.e., standards), and non-typical ones. While the former are available in the professional literature, the latter are usually only gained through years of personal experience. This trait was used in resolving the above contradiction.

Having two groups of associates/consultants was suggested. The first group is composed of entry level employees who are sufficiently proficient in the fundamentals (i.e., typical cases) of both actuarial and investments businesses. These employees can adequately respond to the basic needs of individual clients. To further reduce costs, this group can even be supplied with a computerized database of routine inquires. If the client has a non-typical inquiry requiring more than "text-book" knowledge of the substance matter, then he/she gets help from a small group of highly qualified (and well paid) experts specializing in one of the fields.

A4.3   Summary

The TRIZ methodology, while developed for treating engineering design and manufacturing problems, is applicable to a wide range of management problems. Use of both problem solving methods and the laws of evolution has significant potential for the prompt development of breakthrough solutions to management challenges either directly, or by solving the underlying engineering problems. A close interaction between TRIZ practitioners and representatives of the management profession is called for.

Appendix 5

Glossary[1]

Algorithm for inventive problem solving, ARIZ

The central analytical tool of TRIZ (ARIZ is a Russian abbreviation). Its basis is a sequence of logical procedures to analyze a vague or ill-defined initial problem/situation and transform it into a distinct system conflict. Consideration of the system conflict leads to the formulation of a physical contradiction whose elimination is provided with the help of the separation principles, and by the maximal utilization of the resources of the subject system. ARIZ is a system of the most fundamental concepts and methods of TRIZ, such as ideal technological system (ideal system), system conflict, physical contradiction, the sufield analysis, the Standards, and the laws of technological system evolution.

Altshuller's metrics

See *Technology assessment curves*.

Auxiliary function

A function supporting the system's primary function.

Auxiliary tool

A tool supporting the performance of the main tool(s). Particularly, auxiliary tools perform the function of measurement and/or detection in a system whose primary function is not measurement or detection.

Bi-system

A system consisting of two mono-systems.

Chain sufield

A sufield in which S_2 is controlled by another substance, S_3 (*see* Chapter 3, Table 3.3).

Coefficient of convolution, C_c

A measure of the system's degree of ideality; the ratio of the number of sufields to the number of elements these sufields contain (or the ratio of the number of functions to the number of sufield elements involved in the performance of these functions). For an

---

[1] Alternative terms used by other authors are given in parenthesis.

elementary sufield, $C_c = 1/3$; for a chain sufield, $C_c = 2/5$; for a double sufield, $C_c = 1/2$.

Completely convoluted bi- or poly-system

A completely integrated bi- or poly-system performing two or more functions. In such systems, sub-systems responsible for individual functions are often merged into one substance; their separation is impossible without disintegrating the whole system (e.g., in photochromic reading glasses, two functions – eyesight enhancement and shielding sun light – are performed by one substance – the lens material).

Conflict domain

In ARIZ, the space in the system that contains conflicting components.

Conflicting components

The system's components involved in a system conflict.

Convolution

An evolutionary process of increasing the degree of ideality by eliminating some sub-systems and assigning their functions to other sub-systems.

Degree of ideality

A measure of the system's ideality. This is usually qualitatively expressed as the ratio of the system's functionality over the system's cost and the sum of its problems.

Double sufield

A sufield in which S_2 and/or S_1 are controlled by a second field, F_2 (*see* Chapter 3, Table 3.3).

Effect

The result of the interaction of certain fields and substances. An individual effect can be modeled as shown in Fig. A5.1.

Elementary sufield

A sufield containing two substances and a field (Fig. A5.2).

Environment

The immediate physical surroundings of a technological system or of its part.

Environmental element

In ARIZ, a component that belongs to the environment or an adjacent system and that adversely affects the system's components (e.g., oxygen in the air that causes undesirable oxidation).

Field

The energy needed for the interaction of two substances. In addition to four fundamental fields – electromagnetic, gravitational, and nuclear fields of weak and strong

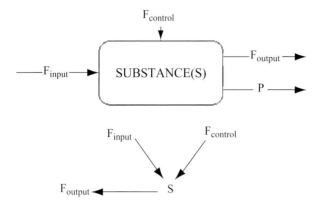

Fig. A5.1. Suffield model of a physical effect.

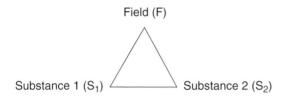

Fig. A5.2. Elementary sufield.

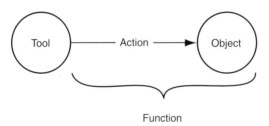

Fig. A5.3. Function.

interactions – TRIZ deals with engineering fields, such as mechanical, thermal, electric, magnetic, and chemical. These fields manifest themselves through many groups of physical and chemical phenomena.

Function

A purposeful physical interaction between two components of a technological system. The description of a function includes the names of both the physical action and the object of that action (Fig. A5.3).

Generic approaches to resolving system conflicts

These are.

- Eliminating either the tool or the object (using the ideality concept).
- Changing the conflicting components (i.e., the tool or the object) such that the harmful action disappears.
- Introducing a special component (tool), intended to eliminate or neutralize the harmful action.

Harmful action

An action that hinders the performance of the primary function.

Heterogeneous bi(poly)-system

A system consisting of mono-systems performing different functions or possessing different physical properties.

Higher-level system (supersystem)

A system that includes the system under consideration as a sub-system.

Homogeneous bi(poly)-system

A system consisting of mono-systems performing similar or identical functions.

Ideal final result

An ideal solution of an engineering design problem based on the notion of the ideal technological system.

Ideal technological system

A system that is absent as a physical entity, but that fully performs the prescribed function.

Ideality tactics

These are generic approaches for realizing the concept of the ideal technological system.

Ideality tactic 1: The object is eliminated, which causes the elimination of the tool and of all its useful and harmful actions.

Ideality tactic 2: The tool is eliminated, and the object (or one of its parts) itself performs the useful action.

Ideality tactic 3: The tool is eliminated, and its useful action is delegated to the environment or to another tool.

Incomplete sufield

A sufield containing fewer than three elements.

Initial situation (initial problem)

Original problem statement, usually a cluster of various problems.

Inventive principles

Techniques for resolving system conflicts used in conjunction with the Matrix.

Inverse bi(poly)-system
A system whose sub-systems have opposite properties.

Law of completeness
This law states that an autonomous technological system must include four minimally functioning principal parts: an engine, a transmission, a working means and a control means.

Law of harmonization
This law states that the necessary condition for the existence of an effective technological system is coordination of periodicity of actions (or natural frequencies) of its parts.

Law of increasing controllability
This law states that technological systems evolve in the direction of increased controllability of their components; this is often achieved by transition from elementary sufields to double and chain sufields.

Law of increasing degree of ideality
This is the primary law of evolution of technological systems. It states that technological systems evolve in the direction of increasing their degree of ideality.

Law of increasing dynamism
This law states that technological systems evolve in a direction toward more flexible structures capable of adapting to changing environmental conditions (multi-functionality) and to varying performance regimes.

Law of non-uniform evolution of sub-systems
This law states that different sub-systems of technological systems evolve at different rates (along their own S-curves); this causes the development of system conflicts.

Law of shortening of energy flow path
This law states that technological systems evolve in the direction of shortening of energy passage through the system (from the engine to the working means).

Laws of technological system evolution
These laws reflect significant, stable and repeatable interactions between elements of technological systems, and between the systems and their environments in the process of evolution.

Law of transition to a higher-level system (supersystem)
This law states that technological systems evolve in the general direction from mono-systems to bi- and poly-systems.

Law of transition to a micro-level

This law states that technological systems evolve in the general direction of the fragmentation of their components (firstly, fragmentation of working means).

Level of invention

A qualitative measure of the degree of novelty of an invention.

Lines of evolution

These lines identify the specific stages of evolution associated with particular laws of technological system evolution.

Macro physical contradiction

This is a physical contradiction formulated at the level of the whole component (e.g., the rod must be hot and cold).

Main tool

A tool performing the primary function.

Matrix, the (Contradiction matrix, Altshuller's matrix)

This is a table that directs the use of the inventive principles for resolving the typical system conflicts.

Maxi-problem

A problem associated with major modifications of a system, i.e., with changing its physical principle of functioning.

Micro physical contradiction

This is a physical contradiction formulated for the components' ingredients (particles), e.g., for the rod to be both hot and cold, its particles must be moving both quickly and slowly.

Minimal technological system

A system consisting of an object, tool, and the energy of their interaction. It can be modeled by an elementary sufield.

Mini-problem

A problem formulated according to the rule: "The system remains unchanged, or even simplifies, but the harmful effect disappears, or the required useful effect is obtained." When solving a mini-problem, the physical principle of the system's functioning is not changed.

Mono-system

A system performing one function.

Object (article, product)
A component of the system that is to be controlled, processed or modified (e.g., moved, machined, bent, turned, heated, expanded, charged, illuminated, measured, detected, etc.).

Partially convoluted bi(poly)-system
A bi- or poly-system with a reduced number of auxiliary components.

Physical action
A physical mechanism that enables the performance of a specific function. For example, a function "cleaning a chemical solution from contaminants" may be based on such diverse physical actions as "moving the contaminants away from the solution," or "disintegration of the contaminants," and/or others.

Physical contradiction
A situation where the same component must satisfy mutually exclusive demands to its physical state (e.g., be hot and cold, electrically conductive and insulative, etc.).

Poly-system
A system consisting of more than two mono-systems.

Primary function
The main purpose of the existence of a technological system.

Psychological inertia
Predilection toward conventional ways to analyze and solve problems.

Resources
Substances, fields and other attributes of a technological system (e.g., time of functioning, occupied space, etc.), as well as of its environment and of an overall system that can be utilized to improve the system.

S-curve
The evolution of technological systems can be illustrated by an S-shaped curve reflecting changes of the system's main performance characteristics (or its benefit-to-cost ratio, degree of ideality) and the time since its inception.

Separation principles
These are generic approaches to resolving physical contradictions: separation of opposite properties in space, separation of opposite properties in time, separation of opposite properties between the whole system and its components.

Shifted bi(poly)-systems (biased bi(poly)-systems)
Bi- or poly-systems whose functionally identical sub-systems differ in certain parameters such as size, color, weight, electrical conductivity, etc.

Standard approaches to solving problems, Standards

A set of the most effective typical transformations of technological systems based on the laws of technological system evolution. Many Standards are written in the substance–field language.

Substance

In the substance–field analysis: an element of a sufield, a technological system of any degree of complexity participating in the performance of a function (physical action).

Substance–field analysis (S-Field analysis)

A branch of TRIZ studying the transformation and evolution of sufield structures.

Sufield (S-Field)

A model of a technological system consisting of substances and fields.

System conflict (engineering contradiction, technical contradiction)

A situation when a useful action simultaneously causes a harmful effect, or the intro-duction (intensification) of the useful action, or elimination (alleviation) of the harmful action, causes an inadequacy or an unacceptable complication of either one part or of the whole system.

System conflict diagram

A graphical representation of a system conflict.

Technology assessment curves, Altshuller metrics

These curves ("number of inventions vs. time," "level of inventions vs. time," "pro-fitability of inventions vs. time") are used to define the position of a system on its S-curve.

Tool

A component having a physical interaction with an object (i.e., controlling the object).

TechNav, guided technology evolution

A systematic TRIZ approach to conceptual development of next-generation products and processes.

Typical system conflicts

Recurring combinations of specific useful and harmful actions.

Useful action

An action that contributes to the performance of the primary function.

Void

A void is a discontinuity in a substance. The void is an exceptional resource because it is always available, extremely cheap and can be easily mixed with other resources forming, for instance, hollow and porous structures, foam, bubbles, etc.

References

Altshuller, G. S. (1963). How to solve scientific problems, preprint, Baku (in Russian).

Altshuller, G. S. (1984). *To Find an Idea*. Novosibirsk: Nauka, p. 69 (in Russian).

Altshuller, G. S. (1988a). Small unbounded worlds. In: *A Thread in the Labyrinth*. Petrozavodsk: Karelia, pp. 165–230 (in Russian).

Altshuller, G. S. (1988b). *Creativity as an Exact Science*. New York: Gordon and Breach.

Altshuller, G. S. (1994). *And Suddenly the Inventor Appeared*. Worcester, MA: Technical Innovation Center.

Altshuller, G. S. (1997). *40 Principles: TRIZ Keys to Technical Innovation*. Worcester, MA: Technical Innovation Center.

Altshuller, G. S. (1999). *The Innovation Algorithm*. Worcester, MA: The Altshuller Institute for TRIZ Studies.

Altshuller, G. S., Shapiro, R. B. (1956). On the psychology of engineering creativity. *Problems of Psychology*, **6**, 37–49 (in Russian).

Altshuller, G. S., Vertkin, I. M. (1994). *How to Become a Genius: The Life Strategy of a Creative Person*. Minsk: Belarus (in Russian).

Altshuller, G., Gadzhiev, Ch., Flikshtein, I. (1973). *Introduction to Sufield Analysis*. Baku, USSR: The Public Laboratory for the Theory of Invention (in Russian).

Astashev, V. K., Babitsky, V. I. (1998). Ultrasonic cutting as a nonlinear (vibro-impact) process. *Ultrasonics*, **36**, 89–96.

Banks, H. T., Smith, R. C., Wang, Y. (1996). *Smart Material Structures: Modeling, Estimation and Control*. New York: John Wiley & Sons.

Bassala, G. (1988). *The Evolution of Technology*. Cambridge: Cambridge University Press.

Brown, S., Rose, J. (1996). FPGA and CPLD architecture: a tutorial. *IEEE Design and Test of Computers*, Summer 1996, 42–57.

Cole, W. (2002). Let's twist again! Technology that enables wing 'warping' rolled out at Dryden. *Boeing Frontiers Online*, Volume 1, Issue 1, May 2002, http://www.boeing.com/news/frontiers/archive/2002/may.

Cusumano, M. A., Markides, C. C. (eds.) (2001). *Strategic Thinking for the Next Economy*. Boston: Jossey-Bass, p. 159.

Davis, T., Sasser, M. (1995). Postponing product differentiation. *Mechanical Engineering*, No. 11, pp. 105–107.

Drexler, K. E. (1992). *Nanosystems: Molecular Machinery, Manufacturing, and Computation*. New York: John Wiley & Sons.

Edwards, B., Cox, C. (2002). Revolutionary concepts for helicopter noise reduction – S.I.L.E.N.T. Program, in *NASA/CR-2002-211650*, May 2002.

Fey, V. R. (1988). *Chronokinematics of Technological Systems*. Baku, USSR: The Laboratory for the Theory of Invention.

Fey, V., Bodine, N., Rivin, E. (2001). A strategy for effective technology investment. http://trizgroup.com/publications/articles/strategy.pdf

Feynman, R. P. (1959). There's plenty of room at the bottom: an invitation to enter a new field of physics. http://www.zyvex.com/nanotech/feynman.html.

Field, F. R., Clark, J. P. (1997). A practical road to lightweight cars, *MIT Technology Review*, September.

Filkovsky, G. L. (1974). *A Scientific Theory as a Machine,* preprint, Baku (in Russian).

Forbes, S. (1997). Fact and comment: plane to see. *Forbes Magazine*, March 10.

Glasmeier, A. (1997). Technological discontinuities and flexible production networks: the case of Switzerland and the world watch industry. In: *Managing Strategic Innovation and Change*. Oxford: Oxford University Press, pp. 32–33.

Jain, V. K. (2002). Advanced fine finishing processes. http://www.geocities.com/aimtdr20th/ keynote/jain.PDF

Kada, M., Smith, L. (2000). Advancements in stacked chip scale packaging (S-CSP) provide system-in-a-package functionality for wireless and handheld applications. In: *Proceedings of Pan Pacific Microelectronics Symposium Conference*.

Leifer, R., McDermott, C. M., O'Connor, G. C., *et al.* (2000). *Radical Innovation: How Mature Companies Can Outsmart Upstarts*. Boston: Harvard Business School Press.

Lown, B. (1996), *The Lost Art of Healing*. Boston-N.Y.: Houghton Miffin Co.

Lynn, G. S., Morone, J. G., Paulson, A. S. (1997). Marketing and discontinuous innovation: the probe and learn process. In: *Managing Strategic Innovation and Change*. New York: Oxford University Press, p. 357.

Malinovsky, G. T. (1988). *Oil-based Coolants.* Moscow, Russia: Chemistry (in Russian).

Mann, D. (2002). *Hands-on Systematic Innovation*. Belgium: CREAX Press Leper.

McGrath, C., McCormick, D. (2002). *The Ultimate Golf Book: A History and a Celebration of the World's Greatest Game*. New York: Houghton Mifflin Company.

McGraw, D. (1996). Staying loose in a tense tech market. *U.S. News and World Report*, July 8, p. 46.

McNeil, B. (2002). Manual windshield wiper conversion and repair. http://www.film.queensu.ca/ CJ3B/Tech/WiperManual.html.

Merkle, R. C. (1997). It's a small, small, small world. *MIT Technology Review,* February/March, pp. 25–32.

Moisan, M., Shivarova, A., Trivelpiece, A. W. (1982). Experimental investigations of the propagation of surface waves along a plasma column. *Plasma Physics*, **24**, No. 11, 1331–1400.

Nolas, G. S., Sharp, J., Goldsmid, H. J. (2001). *Thermoelectrics: Basic Principles and New Materials Developments*. New York: Springer-Verlag.

Papazian J. (2002). Tools of change. http://www.memagazine.org/backissues/feb02/features/ toolsof/toolsod.html

Penzias, A. A. (1997). The next fifty years: some likely impacts of solid-state technology. *Bell Labs Technical Journal*, Autumn, p. 156.

Rantanen, K., Domb, E. (2002). *Simplified TRIZ: New Problem-Solving Applications for Engineers and Manufacturing Professionals*. Boca Raton, FL: St. Lucie Press.

Raskin, A. (2003). A higher plane of problem solving. *Business, ***2.0**, June, pp. 54–57.

Rediniotis, O. I., Wilson, L. N., Lagoudas, D. C., Khan, M. M. (2002). Development of a shape-memory-alloy actuated biomimetic hydrofoil. *Journal of Intelligent Material Systems and Structures,* **13** (January), 35–49.

Rivin, E. I. (1998). *Mechanical Design of Robots,* New York: McGraw-Hill.

Rivin, E. I. (1999). *Stiffness and Damping in Mechanical Design.* New York: Marcel Dekker, pp. 52–86.

Talanquer, V. (2002). Nucleation in gas–liquid transitions. *Journal of Chemical Education,* **79**, No. 7, 877–83.

Tate, K., Domb, E. (1997). *40 Inventive Principles with Examples.* http://www.triz-journal.com/archives/1997/07/b/index.html.

Taylor, M. B., Kim, J., Miller, J., *et al.* (2002). The raw microprocessor: a computational fabric for software circuits and general purpose programs. *IEEE Micro*, Mar–April, pp. 25–35.

Travers, B. (ed.) (1995). *World of Invention.* Farmington Hills, MI: Gale, p. 635.

Tsugawa, S. (2000). An introduction to Demo 2000: the cooperative driving scenario. *IEEE Intelligent Systems*, **15**, No. 4, 78–9.

Vertkin, I. M. (1984). *Mechanisms of Convolution of Technological Systems.* Baku, USSR: The Public Laboratory for the Theory of Invention (in Russian).

Volti, R. (1996). A century of automobility. *Technology and Culture,* **37**, No. 4, 663–85.

Walczyk, D. F., Hardt, D. E. (1998). Design and analysis of reconfigurable discrete dies for sheet metal forming. *SME Journal of Manufacturing Systems,* **17**, No. 6, 436–454.

Wlezien, R. W., Horner, G. C., McGowan, A. R., *et al.* (1998). The aircraft morphing program. In: *Proceedings of the SPIE Smart Structures and Materials Symposium,* Industrial and Commercial Applications Conference, Vol. 3326, paper 3326–20.

Index